Hydrogeochemistry Fundamentals and Advances

Scrivener Publishing
100 Cummings Center, Suite 541J
Beverly, MA 01915-6106
Publishers at Scrivener
Martin Scrivener(martin@scrivenerpublishing.com)
Phillip Carmical (pcarmical@scrivenerpublishing.com)

Hydrogeochemistry Fundamentals and Advances

Volume 1: Groundwater Composition and Chemistry

Viatcheslav V. Tikhomirov

Scrivener Publishing

WILEY

Co-published by John Wiley & Sons, Inc. Hoboken, New Jersey, and Scrivener Publishing LLC, Salem, Massachusetts.
Published simultaneously in Canada.

Cover design by Kris Hackerott

Library of Congress Cataloging-in-Publication Data:

ISBN 978-1-119-16039-7

Printed in the United States of America

10 9 8 7 6 5 4 3 2 1

My loved women,
to my mother, wife and daughter
dedicated!

Contents

Preface

This textbook includes main sections of hydrogeochemistry, methods of its study, terminology and concepts. The textbook is based on the experience and traditions of teaching hydrogeochemistry at the Hydrogeology department of the Sankt-Peterburg State University. These traditions were laid by a brilliant lecturer and scientist Vera Sergeyevna Samarina who taught hydrogeochemistry over a period of almost 40 years and wrote one of the first textbooks in this discipline. These traditions were extended by M.A. Martynova, E.V. Chasovnikova, M.V. Charykova and other lecturers in the department.

The textbook includes three sections. In the first section study methods are reviewed of the geologic medium's hydrogeochemical state. Provided in the section are concepts of analytical ground water composition and properties, and methods of their study. At the conclusion of the section are analysis methods of collected materials, methods of constructing maps, cross-sections and models of ground water geochemical state. The second section introduces spontaneous processes in the water connected with the disruption of thermodynamical equilibrium. The processes are reviewed in consideration of a complex geologic environment, in order to give the idea of methods used for their numerical modeling. The last section reviews external factors of the formation of ground water composition in different climatic and geologic conditions. The spotlight of the section is on the formation of the ground waters' composition, their interaction between themselves and with enclosing rocks. Figuratively, if we view the ground water as a living organism, the first section is discussing its anatomy, the second, its psychology and physiology and the third one, its destiny.

As Hilbert Newton Lewis wrote in the foreword to his *Chemical thermodynamics*, "...a textbook is sort of a restaurant where one can stay his/her hunger without thinking about complex and meticulous processes forming the raw products..." This work is exactly such a textbook and does not pretend to argue controversial hydrogeochemical issues. The main objective of the textbook is to serve previously prepared courses in due order, maximum catchy and gustable. For this reason, the main effort was not the search after truth but systematization and presenting already established provisions.

The publication of this textbook was made possible due to the help by all members of the hydrogeology department of the Geologic faculty at the Sankt-Peterburg State University. I am especially indebted to the department head P.K. Konasavsky and to A.A. Potapov who took upon himself the ungrateful labor of reviewing. I would like also to express my sincere gratitude for the advice, help and useful critique to M.A. Martynova and A.A. Schwartz.

Introduction

Hydrogeochemistry is a science of ground water composition and properties. It studies the distribution of ground water of different properties and composition in the conditions of geologic medium, as well as causes and effects of changes in these properties and composition as they affect the economy. Hydrogeochemistry facilitates the understanding of numerous geologic processes and conditions for the formation of economic deposits, and it solves problems of engineering, geology and ecology. Over time, forecasting and controlling ground water properties and composition has grown more significant in the environment of continuously increasing technogenic effect on nature.

Whereas geochemistry deals with chemical elements' distribution in the composition of the Earth as a whole and hydrochemistry – in the composition of any natural water, hydrogeochemistry concerns the same in the composition only of ground water.

All study methods in hydrogeochemistry lean on the approaches developed in fundamental sciences such as mathematics, chemistry, physics, geology and biology. Thus, to study hydrogeochemistry one needs to have deep knowledge in the basics of these sciences, in particular thermodynamics, chemistry and in recent times also mathematic modeling.

Hydrogeochemistry as an applied science acquired its name relatively late, in the 1920–1930s. Its emergence was caused by the interest to ground waters and by progress in analytical chemistry, which enabled distinguishing ground waters by the composition. Currently hydrochemistry is a scientific discipline of a great practical value. It provides the knowledge necessary for solving problems in lithology, geochemistry, mineralogy, geophysics, exploration for economic deposits, engineering geology and ecology.

HYDROGEOCHEMISTRY: PREHISTORY AND HISTORY

The emergence of hydrogeochemistry as a science was preceded by a thousand-year long prehistory when the concepts of substance of the water, its properties and composition formed. These concepts, similar in appearance, but different in the taste, color and smell of ground waters had been developing way before the emergence of fundamental sectoral sciences (physics, chemistry, geology, etc.)

This prehistory may be broken down into three basic stages: I. pre-Aristotelian, II. Aristotelian and III. post- Aristotelian.

I. The first stage comprises tens of millennia in the human history and ends up with the emergence of ancient natural philosophy. At this stage the water was treated as an animate subject, as deity, and as a terrible element. For this reason the interrelations with the water were initially greater than of a moral nature as with a living being. Only by the very end of this stage was the water treated as an object – the substance with its inherent properties.

People always used water. From the time immemorial they knew that not any water is suitable for their existence, and they knew how to discern it by the quality. So, many applied problems of the present-day hydrogeochemistry were important and were being solved way before its emergence. Most pressing of these issues was undoubtedly the search of waters suitable for drinking, therapy, livestock watering, and irrigation. These qualities were determined sensorily, i.e., by the appearance, smell and taste. Already at that time the people distinguished among natural waters fresh (the Slavs called it *nonfermented, fresh*; the British, *fresh* and the Germans, *frisch*), sour (in English; in German, *sauer*), sweet (Slavs: *sladka voda, slodko woda*; British: *sweet*; German: *suss*; French: *sucre*), salt (English: *salt;* German: *salzig*; French: *sale*) and bitter (Slavs: nasty, bad, worse; British: *bitter*), etc. Such separation of the ground water by taste parameters may be considered to have been the oldest hydrochemical classification. In the areas where there were no fresh-water rivers or lakes the people were looking not just for the water, but for a fresh, sweet ground water. The experience of looking for such waters and improving their qualities was undoubtedly transferred from one generation to the next and accumulated. This experience was valued especially highly in deserts or steppes. This is indicated by the Biblical lore of Moses' miracles during the 40 year-long exodus of the Israelis from the land of Egypt through parched deserts of the Sinai Peninsula. "*22:* So Moses brought Israel from the Red sea, and they went out into the wilderness of Shur; and they went three days in the wilderness, and found no water. *23:* And when they came to Marah, they could not drink of the waters of Marah, for they were bitter: therefore the name of it was called Marah[1]." On advice from the Lord, Moses added wood into the water, and the water became sweet. A salt water spring under this name is known currently on the western shore of the Sinai. The Arabs call it Ayun Musa, i.e., the spring of Moses. Its water has a bitter after-taste due to the elevated content of calcium and potassium sulphate. It may be assumed that Moses threw in the water branches of the elvah shrub, which was growing in the Sinai Desert. These branches contain a lot of oxalic acid, which removes the calcium and potassium sulphate from the water.

In those times, almost any liquid was called water, any gas was called air, and any solid substance was called earth. Fire was the most efficient means of primeval chemical analysis. Whatever burnt seemed as if it was turning into air and earth. This formed the idea of four elements of the universe: earth, water, air and fire. The people still did not see a substantial difference between ice and stone. The word crystal meant ice for the ancient Greeks. Out of the general range of customary matter fell only smelted metals. For this reason the first discovered elements were metals: gold, silver, copper, iron, tin, mercury and lead, which ancient astrologers associated with the sun and major planets.

Eventually the capacity of water to convert to air when heated, or to earth when strongly cooled was noted. Perhaps this exact experience of turning the water into earth and into air became a cause of the water losing its animateness and becoming a substance.

In connection with these, the issues of the essence of water and of the nature of its properties became essential. As described by Classical Greek philosophers, first attempts to answer these questions came down to us by Thales of Miletus (625–547 BC) and Plato of Athens (427–347 BC). They believed in the existence of the primary source of all matter on Earth. This

---

[1] Exodus, Chapter 15, verses 22 and 23 (King James Bible).

philosophical concept, later called ***naive monism,*** became commonly recognized by Classical Greek philosophers. According to this philosophy, the "primary substance" existed, from which emerged all other matter. Thales believed that such primary substance was water.

A different viewpoint on the nature of matter was held by Leucippus of Abdera or Miletus (Vth century BC) and his student Democritus of Abdera (460–370 BC). Contrary to Plato's ideas they rejected infinite divisibility of matter and believed that the water also is composed of an infinite number of indivisible particles of matter - atoms, which are not destroyed and do not emerge. However, these atomistic ideas have been forgotten for two millennia.

A belief existed in those times that land was floating in the ocean and that fresh ground waters in the springs formed from the sea water. Perhaps, this question: how the salt water under the ground converts into a fresh water, began the formation of initial hydrogeochemistry concepts. Plato believed that salinity and bitterness of water simply do not percolate through earth.

II. The second stage covers almost two millennia of indisputable authority of Aristotle (384–322 BC). His influence spread with translation of his works into Syrian, then Arabic and in 12th century into Latin.

Aristotle (384–322 BC)

Aristotle, remaining within the framework of naive monism, attempted to explain element transmutation of one into the other. His doctrine was based on the concepts, not of atoms, but of four pairwise opposite properties whose relative content determined the essence of the elements: humidity and dryness, cold and heat. Variations in quantitative ratios of these properties in the composition of matter determined the transmutation of one substance to the other. The water possesses the largest content of humidity and cold, the air, of heat and humidity. He maintained that different combinations of these properties are responsible for all variety of matter on Earth. Aristotle believed that matter can transmutate into each other, and this capability is due to the existence of some principium (the ether or the fifth essence, *quinta essentia*).

Besides, Aristotle no longer associated the origin of fresh water directly with the sea water. In his belief the underground fresh water formed from the air in cold voids of the Earth. This made him the first one to formulate a very important concept related to the formation of the ground water composition: "Waters are of the same qualities as the earth, through which they flow."

Aristotle's idea that by manipulating the properties, it is possible to convert one substance into another, rendered tremendous influence on the evolution of the natural philosophy and facilitated the emergence of alchemy. In Europe Aristotle's doctrine became popular only in the 12th century, due to the efforts of Albert the Great (around 1193–1280) and Thomas Aquinas (1225–1274). Thus began the Christianization of Aristotle's doctrines and their penetration of Catholic theology. At the same time alchemy became very common in Europe. Its main purpose was finding of the "philosopher's stone" for forming gold, silver, longevity potion, universal solvent, etc. The main means of affecting matter were fire and water. One of the major tenets of alchemy said: "Bodies do not act unless they are dissolved." A consequence was active studies of water properties, its capacity to dissolve other matter and to convert into air and earth. Alchemy achievements facilitated the emergence of metallurgy, glass works, manufacturing of paints and discovery of new elements. However, ideas about the essence of natural water did not change. According to Nicolas Flamel (1330–1417?), alchemists continued to believe that "the dissolution is not the absorption of the bodies by water, but their transmutation or conversion of the bodies into the water, from which they were originally created."

Georgius Bauer (Agricola) (1494–1555) developed the fundamentals of chemical analysis and processing of copper, silver and lead ores. He noted the important role of ground water in ore formation and suggested that ores were "congealed sap of the Earth," i.e., formed from ground water. And in his work "On the place and causes of underground (flows)," published in 1546, he proposed that ground waters formed not only from percolation of rain, river and ocean waters but, very importantly, due to congealing of underground vapors. Herewith he first came up with the idea that the water penetrating deep under the surface could turn into vapor, which rose to the surface, congealed and again formed ground water.

Great Discoveries of early XVIth century facilitated the studies of ground water distribution on Earth and its circulation cycle. In 1569–1580 Jacques Bessonn and Bernard Palissy shaped the modern concept of water circulation cycle on Earth. In 1634 René Descartes (1596–1650) in his "Treatise on light," formulated the concept of Earth's spherical zoning: Earth is composed of flaming liquid core, solid crust and the layers of liquid water and atmosphere. In 1644 C. Claramons made a first estimate of the water amount in the ocean. But the ideas of the nature of water per se practically did not change.

The terms "air" and "vapor" initially were used interchangeably. Galileo Galilei (1564–1642) and René Descartes were among the first who distinguished between them. Jan van Helmont (1579–1644) who introduced the notion of "gas" proposed to consider vapor/steam as a transitional stage of water turning into air.

René Descartes stated that "there is always equal amount of salt" in the sea. His "*Principia Philosophiae,*" published in 1644, included the section "On the nature of water and why it easily converts to the air and to the ice." He tried to explain the transmutation of a fresh water into a salt one by suggesting that it is composed of flexible and rigid particles. If these particles, suitably tied with one another, are separated, some of them (flexible) produce the fresh water and some others (inflexible), the saline water. He assumed that in the process of filtration inflexible particles are retained and the saline sea water becomes fresh. Soon thereafter, in 1674, Robert Boyle (1627–1691) established constancy of the marine water salinity. His determination of the average ocean water salinity differed from the current one just by 1%.

Robert Boyle (1627–1691)

Nevertheless, Jan van Helmont still believed that "all bodies (which considered to be mixed), whatever were their nature, opaque and transparent, solid and liquid, similar and dissimilar (as stones, sulfur, metal, honey, wax, fat, ocher, brain, cartilages, wood, bark, leaves, etc.), are made up actually from the simple water and can be completely converted into a tasteless water, at that not even the smallest fraction of the earthly world will remain".

In the second half of the XVIIth century through the studies of Robert Hooke (1635–1703), Christian Huygens (1629–95), Robert Boyle, Isaac Newton (1643–1727) and others, the boiling temperature of water and melting temperature of ice were determined. In 1772 Jean-André Deluc (1727–1817) found that the water reaches maximum density at a temperature around 4 ^{0}C, and James Watt (1736–1819) forced the steam into working for mankind. Nevertheless, concepts of the nature of water per se practically did not change. And the inventor of a universal steam engine, James Watt, believed that "the air is a modification of the water."

III. The end of domination of Aristotle's ideas was defined by Robert Boyle (1627–1691) when he turned to atomistic ideas of the ancient philosophy as related by Democritus of Abdera. Based on these ideas Robert Boyle created the "corpuscular philosophy" and introduced a concept of the "element" as a minimum indivisible component of any substance, and "chemical analysis." Another large feather in Boyle's cap was the affirmation of a leading role of expertize and experiment as a correctness criterion of any theory. He wrote that, "researchers would render the greatest service to the world if they devoted all their forces to manufacturing experiments, collecting observations and did not establish any theories without preliminarily verifying their veracity through the experiment." His efforts resulted in qualitative change of study techniques. Thereafter chemical experiments were conducted with accurate measuring of the mass of interacting matter. This enabled R. Boyle to prove that fire is not a substance but only a result of burning with the participation of the air.

In the XVIIIth century special attention attracted curative properties of the ground water. Mineral water treatment became fashionable. As the health resort business tempestuously grew, plenty of attention was devoted to the search of mineral ground waters and study of their properties, composition, and formation conditions. In Russia, first scientific interest to mineral waters was associated with the name of Peter the Great. It was he who attracted attention to the need for exploring national natural resources, in particular searching and utilization of curative waters. He also was the originator of first expeditions for the study of Russia's natural treasures and organizer of health resorts on mineral waters. In 1719 first state health resort "Marcial waters" was launched in Karelia. A great role in studies of ground waters in Russia belonged to the Russian Academy of Sciences founded by Peter I and its expeditions for the study of natural treasures in Russia. Ground waters were studied by Stepan Petrovich Krasheninnikov (1711–1755), Ivan Ivanovich Lepekhin (1740–1802), Nikolay Yakovlevich Ozeretskovsky (1750–1827), Nikolay Petrovich Rychkov (1746–1784), Vasily Fedorovich Zuyev (1754–1794), Peter Simon Pallas (1741–1811) and others. Their efforts resulted in the formation in XVIII century of first scientific concepts of ground waters in Russia, which formulated in his works "On layers of Earth" and "On the birth of metals from shaking of Earth," by Mikhail Vasilyevich Lomonosov (1711–1765). In 1785, in France a first *Thesaurus of all mineral springs of the realm with their brief descriptions* was published. But even then, the concepts of the nature of water had hardly changed.

However, measuring the mass of combustion products in the air discovered inexplicable loss of matter. A German physician, Georg Ernst Stahl

(1659–1734) explained this loss by the existence of some matter with negative mass. He named this substance phlogiston. The search of this enigmatic substance had a definitive significance in the evolution of concepts of air composition and facilitated the discovery of hydrogen and oxygen.

Many scientists tried to catch and study this mysterious phlogiston. At last, in 1766 the Englishman, Henry Cavendish (1731–1810) made it. He discovered a substance similar to it. Later this substance, for its exceptional role in the formation of water, was called hydrogen (Latin *Hydrogenium*). Five years later, in 1771, in the work "On the nature of waters" a Frenchman, Antoine Laurent Lavoisier (1743–1794), proved that the water and earth could not convert into each other. The same year, a Swede, Carl Sheele (1742–1786), and in 1774 an Englishman, Joseph Priestley (1733–1804), independently discovered oxygen. They informed A.L. Lavoisier about their discovery, and he found that their substance was a component of the air, acid and many other compounds. In 1777 discoveries of oxygen and nitrogen determined the air composition. These discoveries allowed A.L. Lavoisier to reject the theory of phlogiston and assert the validity of the law of conservation of matter. In 10 years, in 1783–1785 the same indefatigable A. Lavoisier proved that the water was composed of hydrogen and oxygen and cannot convert to the air and back. These successes in chemistry enabled Alexander von Humboldt (1769–1859) and Joseph Louis Gay-Lussac (1778–1850) in 1805 to determine the chemical formula of the solvent in water composition: H_2O.

Antoine Laurent Lavoisier
(1743–1794)

Thus, it was proven that the natural water is a complex solution dominated by the compound of oxygen and hydrogen, H_2O. For this reason further studies of ground water directed to the determination of its dissolved matter were closely associated with successes in chemistry, especially analytical chemistry.

In 1804 John Dalton (1766–1844) published a first table of atomic masses. In 1807–1808 an English physicist, Humphry Davy (1778–1829), discovered sodium and potassium, and he proved the elementary nature of chlorine. The circle of studied atoms rapidly expanded. In 1865 Dmitry Ivanovich Mendeleyev (1834–1907) established periodical law of chemical elements having thereby determined the boundaries of this circle. A little later (in 1896) a French physicist, Antuan Anri Bekkerel (1852–1908), discovered radioactivity, i.e., capacity of some atoms to convert spontaneously into other atoms. This discovery drew attention to radioactive elements, first of all uranium, thorium and radium and products of their decay. By 1911 for 12 places in Mendeleyev's periodic table competed around 40 elements with different radioactive properties. In an attempt to solve this problem, in 1910 Frederic Soddy (1887–1956) came to a conclusion of the existence of elements with similar properties but diffeent atomic mass.

In 1913 he proposed to call such atoms isotopes. Soon thereafter it was proven that beside stable isotopes there may also be radioactive ones. In 1929 William F. Giauque (1895–1982) and a student Garrick Johnston (USA) identified three stable isotopes in the atmospheric oxygen, and in 1932 Harold Clayton Urey (1893–1981) discovered deuterium and heavy water.

At the same time analytical chemistry methods were being developed and improved, which enabled the determination of individual element contents in the composition of various natural matter, including the natural water. The ground water composition was initially studied in order to search for new elements and identify their properties and distribution. Then, early in the XIXth century, appeared the interest to the ground water composition associated with the study of their balneological properties. Gradually the scope of studied ground waters and components in their composition expanded, which facilitated the formation of concepts about the ground water as a composite solution.

In connection with these, at the same time, there appeared theories of the structure and properties of water solutions of electrolytes, and of solution and precipitation processes. The theory of electrolytic dissociation proposed in 1887 by Svante Arrhenius (1859–1927) turned out to be especially fruitful. In 1923 Peter Joseph Debye (1884–1966) and Erich Armand Hückel (1896–1980) proposed a statistical theory of diluted strong electrolytes, which facilitated the transfer from simple concentration of electrolytes to thermodynamic, i.e., to activities.

Simultaneously, in the end of XIX century (1879) a new science formed - hydrogeology, which identifies ground waters as the object of professional attention. Among the problems solved by this science is also the issue of ground water composition and properties. Severe epidemics associated with water-supply (epidemics of the enteric fever in Paris) directed attention in the 1890s to ground water contamination. A result of this was a first service for the sanitary protection of water-supply sources in Paris.

Initially, ground waters were studied within the framework of geochemistry as one of geologic objects. Geochemists soon switched from comparing the composition of individual minerals to comparing composition of rocks, their associations and even entire geospheres. Such comparisons required an assiduous statistical analysis of the distribution of individual chemical elements. An American chemist, Frank Wigglesworth Clarke (1847–1931), expended 40 years of his life dealing with this painstaking and very labor-intensive work. In 1908 he published a fundamental monograph, *The data of geochemistry*, where he included results of his calculations of the Earth crust average elemental composition as well as that of various rocks, ground water, etc. Subsequently these data were numerously fine-tuned by F.W. Clarke himself and by other geochemists. Alexander Yevgenyevich Fersman (1883–1945) proposed to call these average values of individual elements content "clarkes" in honor of F.W. Clarke. Study of the clarke values showed that the element distribution on Earth decreases with the increase of their atomic mass. Greater than over, it so turned out that the contents of isotopes with even sequential numbers were higher than with odd numbers. Subsequent spectral analysis studies of elemental composition in meteorites and star atmospheres showed that these features in Earth composition are common for the galactic cosmic bodies, and that they reflect primordial distribution of elements prescribed by their nuclear properties. Nevertheless, the establishment of geochemistry as a science is associated with the names of Vladimir Ivanovich Vernadsky (1863–1945), A.E. Fersman and Victor Moritz Goldschmidt (1888–1947) who were the first to use the achievements of chemistry and thermodynamics for the explanation of processes within Earth.

Vladimir Ivanovich Vernadsky (1863–1945)

In the spring of 1882 the Russian Geological Committee was formed, where the hydrogeological discipline was overseen by Nikolay Fedorovich Pogrebov (1860–1942), who discovered radon in the waters of lake Lopukhinka. It is reasonable to consider him as a first official Russian hydrogeologist. An American geologist, Chase Palmer (1856–1927), studied waters in oil fields and in 1911 proposed a first ground water classification by the salt composition. This classification was for a long time commonly used abroad and in our country. First systematic ground water and their composition study in Russia is associated with the names of agrologist Vassily Vasilyevich Dokuchayev (1846–1903) and his students. Early in the XXth century he created in Petrograd a chemical laboratory of Russia's Ministry of Agriculture. In 1914 Pavel Vladimirovich Ototsky (1866–1943) noted a regular change in the properties and composition of ground waters in the Russian territory. Chemist Nikolay Semenovich Kurnakov (1860–1941) was among the first who studied brines, muds and salt deposits in Russia, and who introduced the concept of "metamorphization factor," which in 1917 he took as a basis for the classification of salt lakes. The same year J. Rogers observed a change in the composition of waters in California oil fields with depth and made a conclusion about reduction of their sulfates to H_2S. In 1920 analysis of deep ground waters in the US oil fields acquired systematic nature. At the same time in Russia in Novocherkassk by the efforts of Pavel Alexandrovich Kashinsky (1868–1956), who may be considered founding father of the domestic hydrochemistry, and Oleg Alexandrovich Alekin (1908–1995), there was created a first Hydrochemical institute. The studies of this period have been summarized by V.I. Vernadsky in 1929, in the Russian mineralogical society, where he presented a report "On the classification and chemical composition of ground waters." In this report, for the first time, the general discipline was defined, which was named ***geochemistry of ground water*** or ***hydrochemistry***. In 1933–1936 three volumes of *A history of ground waters* were published, in which V.I. Vernadsky systematized and gave an account of whatever was accumulated by the 1930s on this subject. Works by V.I. Vernadsky facilitated the merging of desultory studies on the ground water composition into a single general channel of hydrochemistry.

In 1920 in Novocherkassk the first Hydrochemical institute in the world was created. In the first stage, the underground and surface waters were studied together. In 1938 the term "hydrogeochemistry" appeared, associated with a study of the composition of ground water only. In 1948 O.A. Alekin published a first textbook "General hydrochemistry," in which ground waters were reviewed separately.

In the first stage, the main attention in hydrochemistry was devoted to methods of chemical analysis, and identification of ground waters by the composition, their classification and distribution. Significant attention was allotted to the search and mapping of potable and especially mineral waters and to industrial exploitation of salt lakes. In 1930 at the IVth hydrogeology health-resort conference, a general practitioner and balneologist Mikhail Georgiyevich Kurlov (1859–1932) proposed the formula for a brief and visual description of ground water chemical composition. In 1933 Nestor Ivanovich Tolstikhin (1896–1992) used cyclograms for picturing the ground water composition, which were common until now. In 1935 Mikhail Georgiyevich Valyashko (1907–1984) utilized the schematics of N.S. Kurnakov and proposed his own classification of lake waters by their salt composition. Improvements of ground water hydrochemical classification were performed by Sergey Alexandrovich Shchukarev (1893–1984), Nikolay Nikolayevich Slavyanov (1878–1958), Vladimir Alexeyevich Sulin (1896–1950), O.A. Alekin, A.M. Piper, Alexander Mikhaylovich Ovchinnikov (1904–1969), Otar Sergeyevich Dzikiya, Elena Evstafyevna Belyakova and many others. Currently, practical significance is maintained by M.G. Kurlov's formula, and V.A. Sulin's (1946) and O.A. Alekin's (1948) classifications. In common use abroad is a diagram proposed in 1944 by A.M. Piper, and the pattern presented by Henry A. Stiff in 1951.

Simultaneously, the studies on the distribution of different composition ground waters were published. Russian agrologists were especially interested in ground waters and their composition. In 1923 Vsevolod Sergeyevich Ilyin (1888–1930) proposed a zoning scheme of these waters by the composition. His studies laid the basis for regional hydrogeochemistry in Russia. In 1934–35 the first papers were published by Constantine Lukich Malyarov and V.A. Sulin devoted to waters of Russia's oil fields. Special attention was devoted to mineral waters and lakes. The papers by N.S. Kurakov, N.N. Slavyanov, Vasily Alexandrovich Alexandrov (1877–1956), Alexy Ivanovich Dzens-Litovsky (1892–1971), M.G. Valyashko and others played a great role in this. In 1947 a monograph was published by A.M. Ovchinnikov called *Mineral waters.*

Successes in analytical chemistry enabled substantial expansion of the concepts about composition of ground waters. In 1935–1936 Vasily Petrovich Savchenko (1904–1971) and Anatoly Lvovich Kozlov (1903–1980) noted a correlation between the content of dissolved helium and the age of ground water. This period ended with the appearance of study manuals on the geochemistry of ground waters. In 1949 Victor Alexandrovich Priklonsky (1899–1959) and Fedor Fedorovich Laptev published apparently the first work devoted to a study of ground water composition. In 1953 O.A. Alekin published the first textbook, *Foundations of hydrochemistry.* In 1958 Vera Sergeyevna Samarina (1916–2002) published (apparently) the first textbook in Russia on the ground water composition, *Hydrochemical testing of ground waters.* In the US a similar work by John D. Hem (1916 –1994), *Study and interpretation of ground water chemical parameters* was issued in 1959.

In the second stage, in mid-XXth century, the issues of ground water genesis and formation became greater than pressing. In 1944 Georgy Alexeyevich Maximovich (1904–1979) introduced the concept of "ground water facies," and in 1958 Gregory Nikolayevich Kamensky (1892–1959) turned their attention to genetic types of ground water. Abroad, the first "hydrochemical facies" was identified by William Back (1925–2008) in 1960.

A discussion emerged relatively genesis of the ground waters of different composition, in particular great depth brines. Participants in this discussion were M.G. Valyashko, Alexander Ilyich Perelman (1916–1998), Yefim Vasilyevich Posokhov, I.K. Zaytsev, Yevgeny Victorovich Pinneker (1926–2001), Alla Ivanovna Polivanova (? –1996), Login Nazaryevich Kapchenko (1936–2006) and others. By the end of the 1950s papers appeared devoted directly to biochemical and chemical processes in ground waters. During this period Stanislav Romanovich Kraynov (1928–2007) studied rare elements, Vladimir Mikhaylovich Shvets (b. 1929) - distribution of organic matter and Sergey Ivanovich Kuznetsov (1900 –1987) and Lyudmila Evstafyevna Kramarenko – the role of microorganisms in the formation of the ground water composition. In 1961 Horton B. Craig (1926–2003) established linear correlation between isotope composition of hydrogen and oxygen in meteoric waters and drew attention to isotope composition of ground water.

In solving problems of ground water genesis and formation of its composition, hydrogeochemists turned to the fundamentals of chemical thermodynamics and physical chemistry. A special role in this belongs to the work by Robert Minard Garrels (1916–1988), *Mineral equilibrium at low temperature and pressures*, published in 1960, and in Russia in 1962. His work, especially the monograph, *Solutions, minerals, equilibrium*, written together with Charles Louis Christ (1916–1980), facilitated broader application of the laws and techniques of thermodynamics in studies of ground water formation. Of great significance were the publications of Sergey Alexandrovich Brusilovsky (b. 1936) about migration forms of elements in ground waters and also of Harold S. Helgeson (1931–2007), Igor Konstantinovich Karpov (1932–2005) and Boris Nikolayevich Ryzhenko (b. 1935). In study of brines a great merit belongs to Kenneth Sanborn Pitzer (1914–1997), who proposed in 1973 his high-density brines model. His papers noticeably raised the level of ground water formation hydrochemical studies. They introduced thermodynamical and kinetic methods in studies of the water-rock system, having thereby created the background for hydrochemical modeling. This transition of qualitative analysis of water composition formation processes from empirical to quantitative is the main achievement of this period. It enabled the transition from a description of what was to a prediction of what will be with the ground water composition in specific conditions.

Simultaneously, the field of hydrochemical studies widened. Onland, due to successes in drilling technology, hydrogeochemists penetrated to a depth of 12 km. Studies of mineral and hydrothermal waters became much greater than active. A new discipline appeared in hydrogeochemistry, which studied ground waters of seas and the ocean. This discipline was initially associated with study of ocean-bottom hydrothermal "smokers" and later with the results of deep-water drilling. At the same time, the domain of studied elements and isotopes in ground water expanded. Hessel de Vries (1916–1959), Vassily Ivanovich Ferronsly (b. 1925), Vladimir Timofeyevich Dubinchuk (b. 1936), Vladimir Andreyevich Polyakov and others published papers about natural isotopes of various elements in ground waters.

The third stage in the evolution of hydrogeochemistry was associated with aggravation of the ecological situation in industrial countries. It began in 1980s and was manifested in the interest from hydrogeochemists to investigation of consequences of the anthropogenic ground waters contamination. Successes of hydrogeochemistry in this area were associated in Russia with the names of Valentine Mikhaylovich Goldberg (1934–1996), Faina Ivanovna Tyutyunova and S.R. Kraynov, and in the West with the names of Jean J. Fred, Ch. D. Rail, D.M. Mackay, J.A. Cherry and others.

At this stage a major attention was devoted to technogenic compounds, their penetration subsurface, and spread and interaction when dissolved in ground water with rocks. As a consequence, the domain of hydrochemistry issues of applied nature associated with the determination of not only the state of the medium, but also its changes in real time, noticeably expanded. Successes in computer technology and applied mathematics, which allowed solving the ecological problems, turned out very helpful. In connection with this, the significance of hydrogeochemical monitoring and mathematic modeling drastically increased.

The first hydrogeochemical mathematical model of dissolved substance transfer was proposed, most likely, by J. P. Bredehoeft in 1973. Soon thereafter L.F. Konikov developed a similar model but with inclusion of components' dispersion in the process of migration. In our country, modeling evolved by the efforts of I.K. Karpov, Yuri Vsevolodovich Shvarov, Valery Nikolayevich Ozyabkin (b. 1937), Mikhail Boleslavovich Bukata (1950–2010), Gennady Anatolyevich Solomin, Mikhail V. Mironenko, Marina Valentinovna Charykova (b. 1961) and others. Eventually hydrochemical mathematic modeling became, beside the hydrodynamic one, a most common instrument of the applied hydrogeochemistry.

Numerical hydrochemical modeling stimulated studies in kinetics of the slowest hydrogeochemical processes, in particular, in dissolving speed. Over a relatively brief period, by the efforts of Antonio Lasaga, L. Neil Plummer, E. Lennart Soberg, Victor Alexeyevich Alexeyev (b. 1946) and others, the dissolution velocities of numerous aluminosilicate minerals were researched.

In connection with depletion of reserves and contamination of potable waters, the issue of civil rights for the water became acutely important. A first solution of this problem in Russia became RF law, "On the protection of the natural environment," which was enacted in 1991. Article 85 of the law for the first time in Russia gave the legal determination of ***ecological crime*** as the behavior causing social danger. Objects of this crime became the natural environment and its most significative components, in particular, ground waters. In 1996 ecological crime became the institute of Special part of the Russian criminal law (RF Penal code, Ch. 26: Ecological crimes). Under the Article 250 of this code, "contamination, pollution, depletion of surface or ground water, sources of potable water-supply or other change in their natural properties, if these deeds caused substantial damage to the wildlife or vegetation, fish reserves, forestry or agriculture," are ecological crimes. The consequence of ecological crime is ***ecological liability***, which involves incarceration of up to five years. And in 2007 the ***RF Water Code*** (*ВК РФ*) was enacted, which is a codified regulatory act controlling relations in the sphere of water use in Russia. It includes the entire section (Articles 95, 103 and 104) dealing with the protection of ground water against contamination.

Currently hydrogeochemistry is a fully formed science with a huge army of experts – hydrogeochemists. It is possible to identify within its framework four major disciplines:

1. General Hydrogeochemistry includes these subjects: physicochemical processes in the conditions of geologic medium; their interaction with rocks, oils or subsurface gases as well as migration mobility; balance and circulation of their chemical elements; and formation of ground water composition. Currently working in this discipline, in the Russian Federation, are V.P. Zverev, Klara Yefimovna Pityeva (b. 1924), Alexander Nikolayevich Pavlov (b. 1933), and abroad, Patrick A. Domenik, Franklin V. Schwartz, James I. Driver, and others.
2. Regional hydrogeochemistry studies and maps the distribution and behavior mostly of ground waters of different composition as a result of their formation conditions. Within the framework of this discipline are studied processes of rock weathering, soil formation and ground water composition under the effect of exogenous factors in specific geologic conditions. The whole army of professional hydrogeologists takes part in the description of ground waters in

various geologic conditions and in making hydrogeochemistry maps. In the Russian Federation these are K.E. Pityeva, Vladimir Andreyevich Kiryukhin (1930–2011), Stepan Lvovich Shvartsev (b. 1936) and abroad, James I. Driver, Charles V. Fitter (b. 1941) and others.

3. Endogenous hydrogeochemistry analyzes the formation and evolution of the composition of mostly deep thermal ground waters onland and in the ocean. This discipline studies hydrochemistry and formation of mineral, industrial and thermal waters; formation waters of oil and gas basins; ore and hydrothermal solutions; and their role in formation of economic deposits. In his time V.I. Vernadsky noted the existence of vertical zoning in water composition. Initially, it was studied by Nikolay Klimentyevich Ignatovich (1899–1950), Fedor Alexandrovich Makarenko (1906–1984), Constantine Vasilyevich Filatov (1907–1960s), N.I. Tolstikhin, Ivan Kireyevich Zaytsev (1907–1991), and some others. Currently it is possible to identify two different disciplines in endogenous hydrogeochemistry: oil-gas and hydrothermal. The oil and gas discipline deals with the formation waters of oil and gas basins in an environment of elevated temperatures and pressures, and their part in the formation of oil and gas fields and sedimentary ore bodies. The hydrothermal discipline deals with the formation of ground water composition in conditions of very high temperatures and pressures, and their part in processes of metamorphism and formation of hydrothermal ore deposits.
4. Applied hydrogeochemistry is oriented to the solution of specific economic issues. These issues include:

    a. Search and appraisal of ground water as an economic deposit (potable, mineral, industrial, thermal and other);
    b. Protection and monitoring of ground water quality, especially their commercial development areas;
    c. Engineering evaluation of the interaction between ground water and various technological materials in order to assure the reliability of engineering facilities and underground communications;
    d. Search of ore and oil-gas fields by indirect hydrochemical indications in ground water composition.

The prying spirit of Apostle Thomas
To resurrected saying, "I won't trust it
Until my fingers put inside the wound,"
He prised apart millennia of faith,
He evidences checked against the numbers,
He tint and sound sensed by the touch,
He weighed the light, he measured race of ray,
He brought theology dogmata of the faith
Upon the guise of matter and the forces.

Maximilian Voloshin, 1923.

COMPOSITION AND PROPERTIES OF GROUND WATERS

Before reviewing the composition it is necessary to determine what the underground water is. Concepts about the essence of the notion "underground water" are ambiguous.

Some hydrogeologists include with the underground water any H_2O compound and even its ions H^+ and OH^- in a geologic medium. In such a case natural, water includes steam as well as ice and the water in the composition of minerals below the Earth surface. But steam is a component of the underground gas and migrates together with it. The ice and snow exist only at below freezing point temperatures within frozen ground and are immobile, same as the minerals. And at last H_2O in the composition of minerals is in a bonded state. Mineralogists study this compound as part of a mineral. Besides the compound, H_2O contains only two elements whose mobility is determined by the behavior of those media, which they are components of.

Some other hydrogeologists view ground water only as a liquid water solution, which has taste, smell and complex composition, which may change when affected by natural and anthropogenic factors. This solution is within the geologic medium and is capable of migrating both relative the enclosing rocks and relative the underground gas. This definition of underground water appears to be greater than logical, as only such water that complies with the laws of hydrodynamics is studied by hydrochemists and is capable of being polluted. This water is considered natural underground water, even when recovered on the surface, until its composition is artificially distorted. Even water samples from deep wells are considered to be the samples of natural underground water until the moment of analysis of their properties and composition under normal conditions.

The term "water" is used in two different senses. In its original household meaning it describes a medium, a liquid solution with various gustative qualities, which humans looked for, produced and utilized from the time immemorial. A second meaning appeared only two hundred plus years ago, due to the Lavoisier's discovery. It is used by chemists as applied to the compound H_2O. But they never apply this term to a water solution of variable composition. The compound H_2O in solution may dominate, may yield in weight to the other components (in brines), but it never occurs in the nature in pure form.

Technical engineers prefer using for pure water the term moisture (moisture content, moisture retention, etc.). In order not to say that the water is composed of water and to eliminate ambiguity of this term we will also use for the compound H_2O the term **moisture**. This notion in its household meaning is closest to the notion of the "pure water," i.e., water that is condensed and distilled without dissolved components. For this reason within the framework of this textbook we will treat ***ground water*** (*underground water*) in its original household meaning, i.e., as liquid natural water solution in a geologic medium, and the chemical compound H_2O we will be calling moisture. Moisture as water without dissolved components as the compound H_2O may be a component part in the composition of various natural matter and compounds including in the composition of natural waters, minerals, and underground gas or oil. But the water as solution is not part of the underground gas or mineral. In particular, moisture may be present in magma but not underground water.

1

Analytical Composition and Properties of Ground Water

Underground water as solution includes almost all elements from the Mendeleyev's table. The forces of interatomic and intermolecular interactions form between them chemical bonds varying in strength, which determine the ***real composition*** of the underground water in natural conditions. When the water is recovered onto the surface, taken to a laboratory and prepared to the chemical analysis, some bonds are destroyed on their own and some others are torn apart on purpose. The compounds with the strongest bonds are in effect analyzed, and this does not reflect the genuine composition of waters. For this reason, the composition of waters derived by analysts does not reflect their composition in the natural conditions and is called the ***analytical composition***. As with a nut, in order to find out the water content, it has to be destroyed. Depending on the type of analysis, nature of the effect on the water, and the format of presentation, the resulting following types of analytical compositions are distinguished: the ***isotope***, ***elemental***, ***component*** and ***salt***.

The ***isotope composition*** describes relative content of isotopes in the composition of individual components of a water solution. For instance, it is possible to describe the isotope composition of hydrogen in molecules H_2O, H_2S, CH_4, etc., or of oxygen in compounds H_2O, CO_2, SO_4^{2-} etc., in the composition of underground waters. For this reason, for isotope analysis the studied component often has to be extracted in pure form. The isotope analysis describes only the ratio between contents of isotopes of the same element and is usually expressed by the value of deviation of this ratio from some standard value. The isotope composition of most elements varies within very narrow range. For this reason, it is as a regular rule expressed not in absolute values of the isotope content but in units of deviation of isotope ratios from some ratio accepted as the standard one. This deviation δ is calculated in percent or per mille from the following equation:

$$\delta = 1000 \cdot \left(\frac{R_{\text{sample}}}{R_{\text{st}}} - 1 \right). \tag{1.1}$$

Here, R_{sample} and R_{st} are ratio values usually of less abundant and heavier isotope, and lighter and greater than abundant in the sample and in standard, respectively. The isotope composition of individual elements is among the most conservative composition parameters and changes little in the extraction, testing and analysis of ground water.

Elemental composition of underground water describes relative weight content of individual elements in its composition. For the determination of elemental composition, the water is transferred into atomic state. Almost all (on the order of 85) elements of Mendeleyev's periodic table are found in ground waters. Among them usually dominate O and H. The share of oxygen is 62 to 89% by weight, the share of hydrogen is 8 to 11%. Total share of other elements increases with the growth of water salinity and may reach 30% and higher. Elemental composition of water is also quite conservative and changes only at mass-exchange between the water and other media. It may be substantially distorted in the process of testing and analysis due to the loss of volatile and poorly water-soluble compounds.

Component composition describes the content of individual most stable compounds in the water composition as determined by methods of chemical or physical analyses. The genuine natural composition of ground waters in the process of testing and analysis is distorted, and the distortion is stronger if analysis technique is greater than rigid. As mentioned above, at the strongest action on the water (vaporization, annealing and conversion of dry residue into plasma state) the elemental or isotope composition is determined. At the determination of the component composition, not all, but only the weakest intermolecular and some interatomic bonds are destroyed. For this reason the component composition, which is sometimes called ***gross composition,*** strongly depends not only on testing techniques but also on ground water analysis techniques. Moisture, i.e., H_2O, is only the dominating stable compound in the underground water composition, and plays the role of a major natural solvent. The type of the remaining dissolved components depends on techniques of their analysis. In actuality these molecules are also associated between themselves. According to the analytical methods, all dissolved components are subdivided into: 1. mineral; 2. gas and 3. organic. Beside dissolved components, a great significance in water composition may have suspended matter, both inert and live.

Finally, sometimes water ***salt composition*** is determined, which is the content of dissolved salts and is calculated based on the content of dissolved mineral components.

All our concepts of a genuine natural ground water composition are formed mostly based on their analytical component composition.

1.1 Moisture

The pure distilled water or moisture, according to the common convention, is represented by a relatively simple compound H_2O. It dominates the composition of ground water and to a significant extent determines their properties. These properties depend mostly on the composition and structure of the H_2O molecule and also on the nature of their mutual interaction.

The hydrogen and oxygen isotope composition affects mostly physical properties of the moisture, in particular, it determines its molecular mass.

Hydrogen is represented by three isotopes: protium H, deuterium D and tritium T. Their atomic masses are, respectively, 1.008, 2.014 and 3.016 amu (atomic mass units). The first two isotopes are stable, the last one rapidly decays (half-life period is 12.35 years) with the formation of ^3He and beta-particle:

$$^3H \rightarrow {}^3He + \beta.$$

The stable isotopes are drastically dominated by a lighter protium (H/D=6,700). The deuterium share in moisture composition is on average only 0.0156% of all hydrogen atoms, but this content sometimes declines to 0.0120% and less in thermal underground waters.

Hydrogen isotope composition is usually described by quantitative ratio of deuterium to protium and is measured by the deviation value of this ratio from SMOW (Standard Mean Ocean Water) standard as follows:

$$\delta D = 1000\left(\frac{\left(\frac{D}{H}\right)_{\text{sample}}}{\left(\frac{D}{H}\right)_{\text{SMOW}}} - 1\right). \tag{1.2}$$

As ***SMOW standard*** is used the D/H ratio in the mix of waters collected at depths of 500-2,000 m at a distance from continents in the Pacific, Atlantic, and Indian oceans.

The lightest hydrogen isotopes display the largest mass difference. Deuterium is two times heavier than protium. For this reason even insignificant changes in isotope composition of hydrogen noticeably affect the weight and physical properties of the moisture. Value of δD usually varies between –160‰ and 0‰ in surface waters and between –120‰ and +8‰ underground.

Oxygen is represented by three stable isotopes: ^{16}O, ^{17}O and ^{18}O. Even the lightest isotope, ^{16}O, dominates among them. Its share of all oxygen atoms is 99.759%. The fraction of ^{18}O in natural moisture is only 0.203%. Odd isotopes ^{17}O do not exceed 0.0374% of the total number of oxygen atoms (^{16}O:^{18}O:^{17}O =2,667:5.5:1).

The oxygen isotope composition is usually described by the quantitative ratio deviation of even isotopes $^{18}O/^{16}O$ from their ratio in standarde SMOW:

$$\delta^{18}O = 1000\left[\frac{\left(\frac{^{18}O}{^{16}O}\right)_{\text{sample}}}{\left(\frac{^{18}O}{^{16}O}\right)_{\text{SMOW}}} - 1\right]. \tag{1.3}$$

The $\delta^{18}O$ value is usually between -14‰ and 0‰ in the surface waters and between minus 4‰ and plus 7‰ underground.

Different combinations of hydrogen and oxygen isotopes determine the isotope composition of moisture. The isotopically heaviest molecule $D_2{}^{18}O$ is 22% heavier than the lightest $H_2{}^{16}O$. Isotopically heavy moisture (D_2O) is 10% denser and 23% greater in viscosity than the regular moisture. It boils at 101.42 °C and freezes at +3.8 °C. Such isotopically heavy moisture is present in ground waters in minuscule amounts. The bulk (over 99%) is the protium moisture $H_2{}^{16}O$.

Differences in physical properties of molecules of different isotope composition play an important role in moisture differentiation by the isotope composition and in biologic processes. It was established that isotopically heavy molecules in the moisture act on living organisms negatively and the lighter ones, favorably. For this reason the water with isotopically heavy moisture is sometimes called dead water and with isotopically light moisture, live.

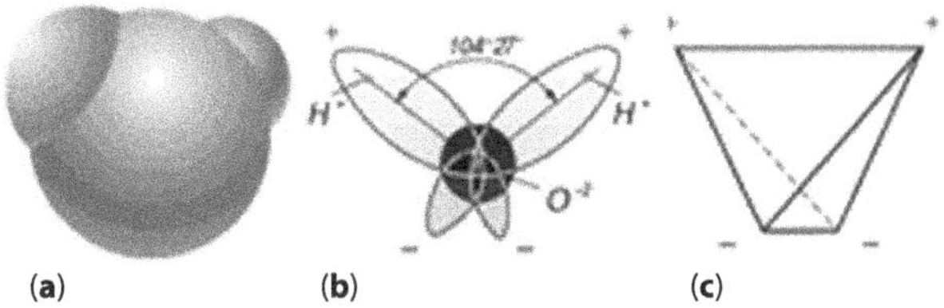

Figure 1.1 Model of H_2O molecules proposed by Niels Bohr: (a) external appearance of its electron cloud; (*b*) positioning of its hydrogen and oxygen atoms; (*c*) positioning of its charges.

The molecule H_2O has a special structure, which provides for its unique chemical properties. In the gas state hydrogen and oxygen atom centers are positioned at a distance of 0.96Å. For this reason, at the diameters of oxygen 1.4Å and hydrogen 1.06Å, water steam molecules form a globe with two small bulges (Figure 1.1, a). And the hydrogen atoms are positioned at the angle not 180° but 104.3° (Figure 1.1, b). For this reason the water molecule forms a dipole (dipole momentum 1.87 Debye) with the positive charge on the side of hydrogen atoms and negative, on the side of oxygen (Figure 1.1, b and c).

Besides, having lost its only electron, hydrogen converts into a small spear with the diameter thousands of times smaller than for the other atoms and with high density of electrostatic stress. Due to this polarized hydrogen, ends of H_2O molecules are capable of coming close to negative charges and form with them an additional weaker bond. This electrostatic attraction (the second positive valence of a hydrogen atom, which it capable of manifesting toward strongly negative atoms) is called ***hydrogen bond***. The length of such bonds between two molecules reaches 3 Å, its energy is almost 19 kj/mole. It is much longer and weaker than interatomic chemical bonds, so it is easily disrupted at the change of environment and in the process of analysis. As a molecule's temperature drops and kinetic energy declines, the influence of hydrogen bonds grows. Whereas at temperature 83°C, average distance between oxygen atoms in the adjacent water molecules is 3.05 Å, and at temperature 15°C, 2.9 Å, in the ice, it is 2.76 Å. At this temperature hydrogen bonds impede mobility of individual H_2O molecules, and thereby facilitate the formation of intermolecular structures. Whereas in vaporific state H_2O molecules do not form structure, in liquid state they join into double, triple and greater than complex associated complexes $(H_2O)_n$. The space between these complexes may be filled up with monomer moisture molecules (Figure 1.2, *b*). But within the temperature interval 0 to 100°C, the concentration of such desultory molecules

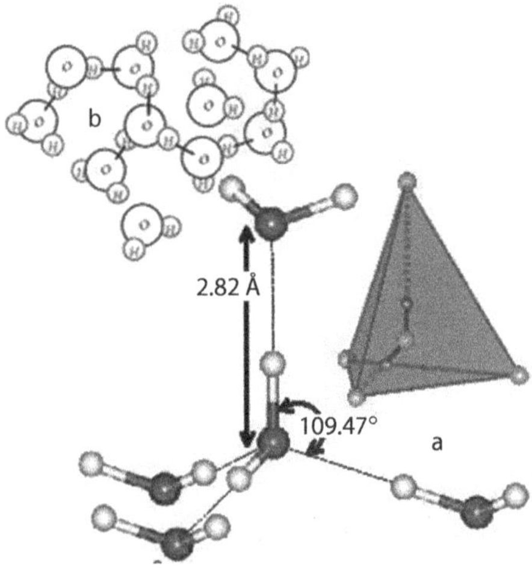

Figure 1.2 Schematics of mutual interaction between H_2O molecules. (a) Tetrahedron of mutual positioning of oxygen atoms in H_2O molecules that associated between themselves. (b) Interaction between H_2O molecules in liquid water. The lines between molecules are hydrogen bonds (White, 1997).

in liquid state is no greater than 1%. In the ice all moisture molecules are associated between themselves and form a structure whose basic element is not exactly regular tetrahedron (Figure 1.2, a). In the center of such tetrahedron is the oxygen atom, and in his apexes are hydrogen atoms. These tetrahedrons together form crystals of hexagonal system with quite loose structureой.

Structure of moisture in liquid state is most complex and not fully understood. It includes also not-tetrahedral structures, some of them of a linear ringlike nature. For this reason also, other hypotheses regarding its structure currently exist but not a single of these hypotheses can convincingly explain all its unique properties. These properties (see Table 1.1) determine a huge and very important role, which natural water played in the formation of Earth as a planet, its stratigraphy, topography, climate and life.

Among these properties must be identified, first of all, the moisture capacity in Earth conditions to exist in three aggregate states: liquid, solid and gaseous (Figure 1.3). This property noticeably separates the oxygen hydride among hydrides of other oxygen group elements (Figure 1.4) and is directly associated with H_2O molecules' capacity to interact between themselves forming complex associations. Almost 98% of all Earth moisture is in liquid state in the waters of the ocean, seas, lakes, rivers and in ground waters. When evaporated, it transits into the atmosphere or underground gas, and when frozen, in glaciers and frozen soils or rocks. The fraction of

Table 1.1 Physicochemical properties of moisture.

Properties		Conditions	Value
Critical temperature			647,3K (374,2°C)
Critical pressure			22,048 MPa
Critical volume of a mole			56,01 $cm^3/mole^{-1}$
Constants of Van der Waals equationя	a		553,23 $kPa\cdot cm^2/mole^{-1}$
	b		30,4 $cm^3 mole^{-1}$
Boiling temperature		101,3 kPa	373,15K (100,0°C)
Melting temperature		101,3 kPa	273,15K (0,00°C)
Max density		3,982°C	1 g/cm^{-3}
Density in liquid state		25°C	0,9971 g/cm^{-3}
Density in solid state		0°C	0,9 g/cm^{-3}
Density in vaporific state		100°C,101,3 kPa	0,00088 g/cm^{-3}
Dynamic viscosity		20 °C	1, 002 mPa·s
Per-unit melting heat		0°C	79,7 $cal\cdot g^{-1}$ (333,69 $J\cdot g^{-1}$)
Per-unit boiling heat		100°C,101,3 kPa	538,9 $cal\cdot g^{-1}$ (2,418 $J\cdot g^{-1}$)
Per-unit heat capacitance in liquid state		15°C, 101,3 kPa	1,00 $cal\cdot g^{-1}.K^{-1}$ (4,1868 $j\cdot g^{-1}.K^{-1}$)
Per-unit heat capacitance in solid state			0,5 $cal\cdot g^{-1}\cdot K^{-1}$ (2,093 $J\cdot g^{-1}.K^{-1}$)
Heat conductivity			0,58 $W\cdot m^{-1}\cdot K^{-1}$
Heat conductivity in solid state		0°C, 101,3 kPa	2,23 $W\cdot m^{-1}\cdot K^{-1}$
Bubble point pressure		20°C, 101,3 kPa	2,337 kPa
Surface tension with air		20°C, 101,3 kPa	0,7275 $mN\cdot cm^{-1}$
Compressibility		20°C, 101,3 kPa	0,47 GPa^{-1}
Heat extension factor		20°C, 101,3 kPa	0,00018 K^{-1}
Relative dielectric permeabilithy		25°C, 101,3 kPa	81,0 un. CGS electrostatic system
Specific electric resistivity		20°C	10^4 ohm·m
Dissociation constant		25°C	10^{-14} $moli^2.l^{-2}$

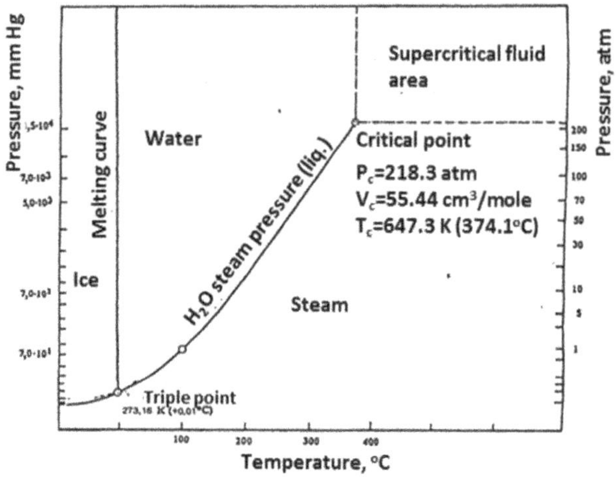

Figure 1.3 H_2O phase state diagram.

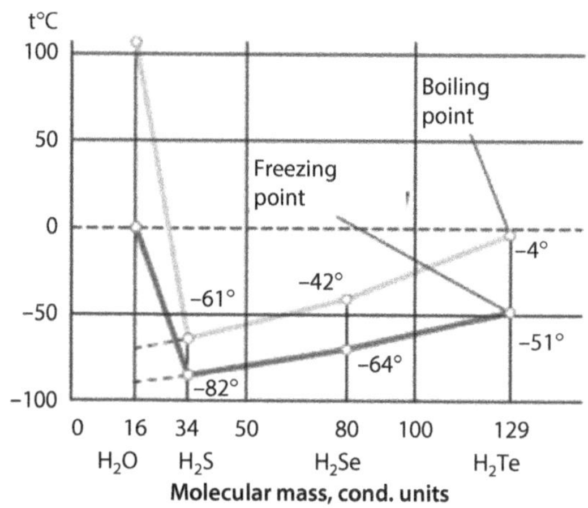

Figure 1.4 Boiling (top curve) and freezing (bottom curve) temperatures of oxide subgroup hydrides of elements.

moisture in the composition of natural gases (first of all in the atmosphere) does not exceed 0.001%, and in the composition of glaciers and frozen soils, 1.8% of the water volume on Earth. Under Earth's temperature conditions moisture is capable of evaporating and condensing. That facilitates the accumulation on the surface of our planet of desalinated waters absolutely necessary for the support of life. Migration of huge ground water amounts causes mechanical and chemical denudation, transport of large mass of suspended solids and dissolved matter, accumulation of deposits, and formation of sedimentary rocks on Earth surface.

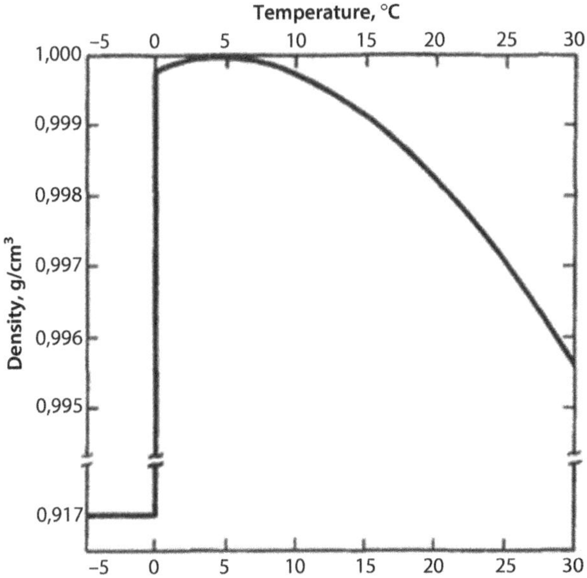

Figure 1.5 Pure water density vs. temperature at pressure 1 atm.

A second important property of the H_2O molecules is associated with their capacity to interact between themselves and form quite loose structures; it is expressed in an unusual dependence of their state on the temperature (Figure 1.4). Moisture has max density of 1.000 g/cm^3 in liquid state at temperature of 4° C. At higher and lower temperature its volume increases and its density noticeably declines. Thus, at the temperature of 25° C its density declines to 0.9971 g/cm^3. But the most surprising is that with its transition from liquid state to solid, the volume increases and the density drops by the whole 10% (Figure 1.5). With this phenomenon is associated also a drastic decline in its heat capacity, almost by half.

Due to these properties the ice floats and protects the rivers and lakes against freezing solid. Near the bottom of rivers and lakes in summer gather the coldest, and in winter, the warmest waters. For this reason, as a regular rule, temperature does not decline below 4°C. At the same time, moisture possesses very high heat capacity, which facilitates heat accumulation in waters and limits the fluctuation of their temperatures. Comparing heat capacities of various matter, (Table 1.2) shows that moisture in liquid state is the most effective heat transfer agent in the lithosphere. This property facilitates the formation of hydrotherms and their use as an energy source.

A third important moisture property is associated with active forces of its intermolecular interaction. The consequence is the greatest among

Table 1.2 Heat capacities, $kcal \cdot kg^{-1} \cdot K^{-1}$.

Substance	Heat capacity	Substance	Heat capacity
Water distilled	1.0	Concrete	0.22
Ice	0.5	Limestone	0.22
Ethyl alkohol	0.57	Glass	0.17–0.19
Pentame	0.52	Granite	0.155
Light gasoline	0.5	Iron	0.111
Benzene	0.322	Copper	0.092

Table 1.3 Surface tension of liquids on contact with air at 20 °C, $mH \cdot cm^{-1}$.

Substance	Surface tension factor	Substance	Surface tension factor
Water	0.73	Oil	0.26
Glycerine	0.59	Ethyl alcohol	0.22
Benzene	0.29	Ethyl ether	0.17
Kerosene	0.29	Mercury	4.86

liquids energy of its surface tension (Table 1.3). The only liquid with higher surface tension is mercury. This property plays an important role in physical interaction between ground water and rocks, oil and subsurface gas.

A fourth important property of moisture is associated with the dipole structure of its molecule. Moisture has the highest value of dielectric permeability among liquids. In pure water, forces of interaction between electric charges are 81 times lower than in vacuum (Table 1.4), which hinders migration of electrons and increases electric resistivity. At multi-step distillation, specific electric conductivity of water declines to 0.8 – 1 $\mu Sm \cdot cm^{-1}$ (specific resistivity about 10^4 ohmm). In its electric properties moisture behaves like a semi-conductor; it facilitates dissociation of electrolytes into ions and prevents their interaction between themselves. Greater than over, due to its capability to dissociate into H^+ and OH^-, it actively participates in acid-alkali reactions. Due to these properties moisture is the most active natural solvent on Earth. It is capable of dissolving solid, gas and liquid matter. There is apparently no matter on Earth, which moisture could not dissolve. Unique properties of H_2O as a solvent, and its vast development on the surface and in the subsurface of Earth, determine the great role of natural water - transporter, which it plays on our planet.

Moisture content in ground water is not analyzed, but it is determined by calculation based on the data on contents of dissolved components. Its content by weight is assumed to be equal to the difference between

Table 1.4 Dielectric permeability (CGSE units) and specific electric resistivity (ohmm).

Substance	Dielectric permeability	Specific resistivity	Substance	Dielectric permeability	Specific resistivity
Water distilled	81	$2 \cdot 10^4$	Ebonite	2.5–4.0	10^{16}
Ethyl alcohol	25.1	$1.5 \cdot 10^5$	Polyethylene	2.3	$10^{10}–10^{13}$
Glass	5–15	$10^{11}–10^{11}$	Air	1.00059	$10^8–10^9$
Porcelain	4.5–6.5	10^{14}	Vacuum	1.0	

underground water's specific gravity and the total content by weight of dissolved components. As the latter noticeably dominate by weight of mineral components, moisture content is usually estimated from the following equation:

$$C_{m,H_2O} \approx 1000\gamma_w - m, \tag{1.4}$$

where C_{m,H_2O} is moisture concentration by weight, $g \cdot l^{-1}$, γ_w is water specific gravity, $g \cdot cm^{-3}$, m is content of water-dissolved components, $g \cdot l^{-1}$.

1.2 Mineral Components

Mineral components include those nonvolatile analyzed compounds, which at oversaturation, as a regular rule, form salts and precipitate. At evaporation of moisture they account for a large part of ***solid residue*** (See Chemical properties). In water solution these components are represented by ions, which react with H_2O and form hydrates. Thereby, moisture hinders ions from reacting between themselves and facilitates an increase in the solubility of their salts.

The number of analyzed mineral components is very significant. They include almost all elements from the Mendeleyev's table. Identification of ground water usually involves a limited number of mineral components, which render maximum influence on the chemical properties of waters. Two main groups are identified among these components by distribution and significance: macro components and micro components.

Macro components include a minimum number of mineral components, without which it is impossible to distinguish waters by the composition. In different countries the set of these components may be slightly different. Nevertheless, they always include cations of sodium Na^+, calcium

Ca^{2+}, magnesium Mg^{2+} and anions of chloride Cl^-, sulphate SO_4^{2-} and hydrocarbonate HCO_3^-. They are called ***major macro components***. Macro components, which are not always analyzed, are called *minor constituents*. This group encompasses K^+, Fe^{3+}, Fe^{2+}, Al^{3+}, CO_3^{2-}, SiO_3^{2-}, and nitrogen compounds NO_2^-, NO_3^-, NH_4^+. Sometimes this group is called mesocomponents. Abroad this series sometimes includes Sr, B, F and Br.

The macro components comprise 90–95% of all matter by weight if dissolved in fresh water and greater than 99% in salt waters and brines. Exactly these components enable distinguishing between various waters by chemical, physical properties and by taste.

Elevated contents of many of them make ground water unsuitable for drinking. Most dangerous among macro components are Br, Fe, B, Sr, NO_2^-, NO_3^- and F. For this reason the determination of macro components' content is necessary almost in all hydrochemical studies.

The remaining analyzed mineral components are called ***micro components*** (*trace constituents*). Their content usually does not exceed 0.1 $mg \cdot l^{-1}$. Some scientists identify ***ultra micro components*** with contents below 0.01 $mg \cdot l^{-1}$. Either of them affect general properties of ground water only slightly but may pertain to their search, balneological, ecological and other qualities. Even at very low concentration they may be life-threatening. Most dangerous among them are Hg, Be, Te, Cd, Se, Pb, Sb, V, As, Cr, cyanides (CN^-) and some others.

The content of almost 50 mineral components in drinking water is regulated by GOST's and SanPiNs[1] (GN 2.1.5.689-98, GN 2.1.5.690-98, SanPiN 2.1.4.544-96, SanPiN 2.1.4.1074-01, SanPiN 2.1.4.1116-02, SanPiN 2.1.4.116-2002).

All mineral components undergo similar methods of study, which include three major stages: 1. Testing and preparation, i.e., preparing samples; 2. Chemical analysis; and 3. Conversion of analysis results into most convenient dimension of concentration.

1.2.1 Testing and Preparation

Testing includes collection of underground water samples and their preparation for safekeeping and chemical analysis. Based on the composition of individual samples, judgment is made about the composition of a huge volume of water in the nature. As a regular rule, several work samples

[1] GOST is State Standard, namely main category standards in the USSR and RF. SanPiN is the regulations establishing criteria for security and safety factors in the environment and to ensure favorable conditions of life for humans.

correspond with each clause of hydrochemical testing, which are called ***spot test samples***. Each of such samples is doubled by 1 or 2 additional samples or accompanied by control samples. Sometimes it is believed to be sufficient to collect ***composite test samples***, which are a mix of several spot samples.

It is practically impossible not to distort natural water composition in the process of its testing, safekeeping, preparation and analysis. Many ions of heavy metals are present only in trace amounts and may not be preserved in the sample due to their oxidizing, reduction, precipitation or coprecipitation, adsorption and ion exchange. In particular, iron is inconvenient in studies due to its tendency to precipitate from a solution at oxidation and to coprecipitate ions of other heavy metals. It is important to monitor very carefully the preservation of its elemental composition, i.e., to eliminate the loss due to the precipitation, degassing or adsorption.

For this reason collection of samples for the study of mineral components in responsible cases is strictly regulated. In Russia the process is controlled under GOST 51592-2000, "Water General requirements to sample collection," which establishes general requirements for the collection, transport, safekeeping and preparation of samples intended for the determination of ground water composition and properties. In the USA these methods are regulated by the Environment Protection Agency (EPA).

Most mineral components are not volatile. When the water is extracted to the surface, tested and transported, they are preserved in dissolved form or a precipitate. For this reason testing for macro components of fresh, brakish and salt water, especially where it is not deep subsurface, does not require special precautions. It is important for the vessel and stopper to be clean and not to introduce the analyzed elements.

Saturated brines may be an exception. They may lose some mineral components to the precipitate due to a decline in temperatures and pressures when water is extracted to the surface. But greater than common loss of mineral components is associated with degassing of CO_2. In gas-saturated waters this loss may be considerable even at depths shallower than 100 m. For this reason when testing deep brines and gas-saturated waters, it is recommended to use special depth samplers.

The preservation of individual components may depend on pH value, content of O_2, degassing or biochemical processes in the sample in safekeeping. If analyses of water composition are conducted not in the field, for preservation of samples, the water is prepared, i.e., conserved or concentrated, and that may cause changes in the content of the analyte.

The conservation may involve filtration, change in pH value, suppression of microorganisms' activity, elimination of a contact with air (with

Table 1.5 Technique of sample preparation for analysis of mineral components.

Operations		Mineral components
Conservation	Filtration (0,45 μm)	Dissolved metals, NH_3+NH_4^+, NO_2^-, NO_3^-, PO_4^{3-}
	Acidifying to pH < 2	Metals, NH_4^+, NO_2^-, NO_3^-, total P, PO_4^{3-}, N_2H_4 (hydrozene)
	Alkalifying to pH > 11	I^-, CN^- (total cyanides)
	Suppression of microorganisms	NH_3+NH_4^+, NO_2^-, NO_3^-, polyphosphates,
	Eliminating contact with air	NO_2^-, NO_3^-, SO_4^{2-}, B, Fe^{2+}
	Cooling to temperatures 2– 5°C	NH_3+NH_4^+, NO_2^-, NO_3^-, BrO_3^-, Br^-, polyphosphates, total P, PO_4^{3-}, SO_4^{2-}, I^-,CN^-,Cr^{6+}
	Freezing to minus 20 °C	NH_3+NH_4^+ (combined), total P, PO_4^{3-}
Dilution	Addition of distilled water	Na^+, Cl^-
Concentrating	Vaporization	Metals
	Sorption	Cations on cation-exchange resins, anions on anion-exchange resins
	Precipitation and coprecipitation	Metals
	Extraction	Metals

O_2), cooling or even freezing he samples (Table 1.5). The conservation methods for cations and anions noticeably differ.

Cations are represented mostly by metals, which are in dissolved and adsorbed state. Adsorbents are solid matter suspended in water, mostly Fe-Mn oxides. In connection with this, total metal content in water is equal to:

$$C_{w,i}=C_{w,i,s}+C_{w,i,ad},$$

where $C_{w,i,s}$ and $C_{w,i,ad}$ are weight contents of the metal i in dissolved and adsorbed state, $C_{w,i}$ is total content of the same metal. Samples for analysis of the summary content of individual metals $C_{w,i}$ are not filtered. For analysis of content of only dissolved metals $C_{w,i,s}$, the samples are mandatorily filtered through a membrane filter with pore size 0.45 μm. Nevertheless, such a filter is capable of getting through well-dispersed Fe and Al hydroxides and even some microorganisms. For this reason sometimes filter 0.1 μm is required.

Metals in water may be components of various compounds. In order to bring them to a single form of existence, the water is acidified to pH below 2. For this purpose acids HCl, HNO_3, H_2SO_4, $HclO_4$ or their combination are used. Acid selection depends on the nature of the metal and analysis technique. For instance, for alkali-earth metals H_2SO_4 should not be used, and in the case of using a graphite kiln for the atom-absorption spectrometry, HCl should not be used. The most convenient is nitric acid – HNO_3. Usually 0.7 ml of 65% HNO_3 is sufficient in order to neutralize the alkalinity and acidify in 100 ml of the sample. The exceptions from the rule are iron and mercury, which require special conservation.

In order to avoid oxidizing of the components by air oxygen, the volume of samples is filled up completely, not leaving place for air. This technique is important to observe in samples for reduced iron.

Anions are represented by compounds greater than diverse in properties. For this reason some of them (halogens, HCO_3^- and SO_4^{2-}) practically do not require conservation, whereas anions with high reducing potential require individual means of stabilizing (S^{2-}, SO_3^{2-}). The contents of associated nitrogen, carbon and phosphorus may change due to the effect of microorganisms. Their conservation is accompanied by addition of antiseptics (chromatophore, timole, etc.). For the conservation of most anions, the samples are filtered through a membrane filter with pore size 0.45 μm, acidified by acids (HCl, HNO_3, H_2SO_4 and CH_3COOH) to pH below 2, isolated from air and cooled to 2 – 5 °C. If acidifying with acetic acid (0.5 ml per 100 ml maximum), it is necessary to add some formaldehyde (0.2 ml per 100 ml) in order to avoid the appearance of mildew (Table 1.5).

Obviously, samples for the content of metals and anions in most cases should be collected separately. At last, most reputable manuals recommend shrinking to the minimum the safekeeping time between the sample collection and analysis. For this reason the time interval of safekeeping water samples must be restricted, and some parameters should be determined directly in the field. The requirements to material of used vessels and stoppers, to conservation methods, and to the duration of safekeeping, are listed in Tables 1–5 and in the attachment.

In recent times requirements to lower boundaries of determined micro component concentrations are becoming greater than rigid. Ever greater than often it is required to determine their concentration up to a nanogram or even ittering per liter. To expand the range of applicability of analytical techniques it is often necessary to increase the concentration of micro components, i.e., to concentrate them. There are several methods of concentrating micro components. The most commonly used is vaporization. This technique allows increasing the concentration of a component 10 to 1,000 times. As a regular rule, solid residue is used for analyzing their content by atomic spectroscopy methods. Sorption is sometimes used for concentrating. For that several liters of water are filtered through ion exchange resins, which are specially prepared adsorbents. Before analysis, the adsorbed ions are displaced by a small volume of hydrochloric acid solution. This way the concentration increases 10 to 100 times. A quite effective concentrating technique is precipitation or coprecipitation when the required analyte is removed from the solution into the precipitate. If it becomes possible to transfer charged ions into the composition of uncharged complex compounds, it is then possible to concentrate it by way of extraction with organic solvent.

1.2.2 Chemical Analysis

Mineral components determine main chemical properties of the ground water, and their content is a major criterion in hydrochemical classifications. Macro components play a leading role among them. Complex analysis of the content of dissolved macro components for the purpose of exploring general ground water properties is called ***routine water analysis***. Analysis of ground water micro components content is ***special water analysis***. Depending on the conditions, all analyses are subdivided into ***field analyses*** and ***laboratory analyses.***

The routine analysis is conducted in any hydrochemical studies as it displays the face of ground water and enables distinguishing it from the other ground waters. No hydrochemical study is possible without the data of the routine analysis. The routine macro components analysis may be complete or shortcut.

The complete analysis includes the determination of main and secondary macro components (Cl^-, SO_4^{2-}, HCO_3^-, Na^+, K^+, Ca^{2+}, Mg^{2+}, CO_3^{2-}, Fe^{+2}, Fe^{+3}, NO_2^-, NO_3^- and NH_4^+). The Na^+ and K^+ content is often determined as a total, $Na^+ + K^+$. The set of analyzed components may vary but always includes main components. Sometimes complete analysis involves the determination of concentrations for some micro components. The complete analysis is conducted in special analytical laboratories, which guarantee the required sensitivity and accuracy.

The shortcut analysis is conducted in laboratory or field conditions and accepts a decreased number of macro components (Cl^-, SO_4^{2-}, HCO_3^- and Ca^{2+}) and sometimes also CO_3^{2-}, Fe^{+2}, Fe^{+3}, NO_2^-, NO_3^- and NH_4^+. The Na^++K^+ and Mg^{2+} contents are often determined by calculation. The field complete analysis is conducted directly at a water show with the help of portable devices or field laboratories. It eliminates safekeeping and transport of water samples and the associated errors. In the laboratory shortcut analysis the samples are studied under normal conditions, either on a field base or in a mobile laboratory.

Special water analysis includes the determination of one or several components, which are not analyzed in the routine analysis, i.e., mostly micro components. This analysis also may be either laboratory or field.

Besides this, all analysis techniques are subdivided into ***regulatory methods*** and ***consensus methods***. The former are developed and recommended as mandatory at the estimation of potable water composition and qualities. In Russia these methods are established and enforced by Russia's Federal Agency for technical regulation and metrology (**Gosstandard RF). In the US these functions are performed by The Environmental Protection Agency of the USA** (U.S. EPA). Consensus techniques comprise all other analysis techniques with inter-laboratory control.

The analytical methods and conditions are selected so that they would provide for error acceptable for the solution of assigned tasks. Standard analysis methods of individual mineral components in mineral-potable, curative-potable or natural potable waters are regulated in Russia by the GOST series 23268. In particular, methods of the determination of some inorganic matter for monitoring qualities of the drinking water are determined by GOST P 51232-98. These techniques are described in greater than details below. In the USA analysis of mineral components is regulated by standard methods of the U.S. EPA, series 200 and 300, intended for the determination of mineral components in municipal potable waters.

1.2.3 Processing of Analysis Results

Laboratories usually release analyses results of mineral components in ***weight concentration***, i.e., in grams or milligrams per unit volum of solution ($\mu g \cdot l^{-1}$, $mg \cdot l^{-1}$, $g \cdot l^{-1}$ or $kg \cdot l^{-1}$). In the USA weight concentrations in *ppm*, i.e., in $mg \cdot kg^{-1}$ are often used. Total content of mineral components in these units is called ***salinity*** (*dissolved solids*).

Weight format of expressing the composition is inconvenient for chemical and hydrochemical studies. What is necessary to compare is not the weight but the number of interacting components. In order to coconsider this number portions are used called moles. ***Moles*** are the mass of $6.02{\cdot}10^{23}$ pieces of any component. This mass, if expressed in grams, is numerically equal to the molecular mass and is called ***molar mass***, and the number of component pieces in it is equal to Amedeo Avogadro's (1776–1856) number. Therefore, equal number of moles of different components means equal number of molecules or ions.

The number of moles of any component per unit volume of natural water is called its ***molarity*** and is determined from the following equation:

$$C_{M,i} = \frac{C_{w,i}}{M_i}, \tag{1.5}$$

where $C_{M,i}$ is molarity of a component i in mole·l^{-1}, $C_{w,i}$ is weight content of the same component in g·l^{-1}, M_i is molecular mass of the same component in g·mole^{-1}.

The most correct expression of concentrations is the ***molar fraction.*** This presents the share of moles of the component i in the sum of moles of all components in the solution. For reliable determination of molar fractions it is necessary to know molarities of all dissolved components, which substantially restricts the use of this dimension. In case of the fresh water the total molar fraction of dissolved components is negligibly small and may be disregarded. Then the molar share of any mineral component i is possible to treat as the ratio of its molarity to the solvent molarity, i.e., H_2O:

$$C_i \approx \frac{C_{M,i}}{\frac{1000\gamma_w - m}{18}} = 18\frac{C_{w,i}}{M_i \cdot (1000\gamma_w - m)}, \tag{1.6}$$

where 18 is the molecular weight of H_2O, g·mole^{-1}, γ_w – specific gravity of water in g·ml^{-1} and m is water salinity in g·l^{-1}. In salt waters where molarities of the dissolved components and of the solvent are quite comparable in value such approximate calculation contains large errors.

The most convenient and common format of expressing concentration of mineral components in hydrochemistry is molality. ***Molality*** of component i is the ratio of the number of its moles to the weight of only solvent, i.e., moisture, in kilograms. Conversion of molarities in molality is possible using the following equation:

$$C_{m,i} = \frac{1000C_{w,i}}{M_i(1000\gamma_w - m)} = \frac{1000C_{M,i}}{1000\gamma_w - m} \approx \frac{1000}{18}C_i = 55.5C_i \tag{1.7}$$

Here, specific gravity γ_w is in g·ml^{-1} and salinity m is in g·l^{-1}. The molality dimension is mole·kg^{-1}.

With the dilution of solutions, concentration of the solvent (1000· γ_w $-m$) tends to 1000·γ_w, i.e., to 1 g·l^{-1}. For this reason at very low salinity

$$C_{m,i} \approx C_{M,i} \approx 55{,}5C_i. \tag{1.8}$$

Thus, with the dilution of a solution the difference between molar and molal concentration vanishes.

Charged ions in electrolytes, as opposed to other components, interact between themselves in proportion not with their number but with their charge. For this reason, for instance, 1 mole of the cation Ca^{2+} interacts with two moles of anion HCO_3^-. For this reason ions are convenient to compare not by the number of molar masses but by the number of their ***equivalent masses*** (molar masses of the equivalent of the substance), i.e., masses with the charge equivalent to the charge of 1 g of hydrogen. The equivalent mass is equal to mole mass of ion divided by its charge. The ion concentration expressed in the number of equivalent masses in 1 liter of a solution is called ***normality*** or ***equivalent form*** of expressing concentrations. Conversion of analysis data from weight units to the equivalent format is done using the following equation:

$$C_{\mathrm{eq},i} = C_{M,i} \cdot z_i = \frac{C_{\mathrm{w},i} \cdot z_i}{M_i}, \tag{1.9}$$

where $C_{\mathrm{eq,i}}$ is ion concentration i in equivalent format; z_i is its charge. In hydrogeology the equivalent format is usually referred to 10^{-3} mole, i.e., to moles in 1 liter of the solution, and denoted (meq·l^{-1}). In the USA equivalent content may be referred to by the weight of the solution in parts per million (ppm), i.e., in meq·kg^{-1} (epm). As the ratio meq·l^{-1} to meq·kg^{-1} describes the solution density, the difference between these equivalent concentrations disappears with the decrease in salinity. In fresh waters they are practically equal.

The equivalent format may be expressed in relative values, as the fraction of the equivalent number of component i in the sum of the equivalent numbers of all mineral components:

$$\left[C_i\right] = 100\frac{C_{\mathrm{eq},i}}{\sum_1^n C_{\mathrm{eq},j}}, \tag{I.10}$$

where $[C_i]$ is percent of the number of ion i equivalent, in equiv-%; $C_{\mathrm{eq},j}$ is equivalent content of components with sequential number j, meq·l^{-1}.

As the natural water is electrically-neutral, the sum of numbers of cation equivalents must be equal to the same sum of anions. The difference between these sums is the criterion of completeness and relablility of the ground water mineral composition analysis. When the analysis is quite complete and reliable, this difference should not exceed 1% of the total content. The difference in excess of ±5% is considered unacceptable.

Electric neutrality of ground water is used also for determination of Na^+ and K^+ content. In shortcut routine analysis these components usually are not analyzed but their total content is calculated as difference between total contents of anions and cations in equivalent format. However, in this case all analysis errors are included in Na^++K^+ content.

Example 1.1 Macro component analysis results in waters of the Czar spring (Karelia) with density 1.013 g·ml^{-1} are given in mg·l^{-1}. These results are represented in column 2 of the following Table. Required: to express these concentrations in units of molarities, molar fraction, molality and in equivalent format.

Let us determine molarity of the components. For instance, for SO_4^{2-} molecular weight M_{SO4}= 32.0+4·16.0 = 96.0 g·mole^{-1}. In connection with these its molar concentration is equal to:

$$C_{M,SO_4} = \frac{35.8\,\mathrm{mg}\cdot\mathrm{l}^{-1}}{96\mathrm{g}\cdot\mathrm{mol}^{-1}} = 0.37\,\mathrm{mmole}\cdot\mathrm{l}^{-1}$$

Molar concentrations of the rest of macro components are listed in the Table below, in column 3. The sum of these molar concentrations is equal to 1.58 mole·l^{-1}.

Molar fraction in the fresh waters is determined as the ratio of the molarity component to total molarity of the solution. Not all components of the solution are analyzed, so the precise value of these concentrations, as a regular rule, is not reachable. For this reason an approximate estimate of molarities is sufficient, assuming total molarity is equal to molarity of the solvent, i.e., H_2O. At salinity 183.6 mg·l^{-1} and water density 1.013 g·ml^{-1} in the case of a sulphate we have:

$$C_{SO_4} = \frac{18}{M_{SO_4}}\cdot\frac{C_{w,SO_4}}{(1000d-m)} = \frac{18}{96}\cdot\frac{0.0358}{(1013-183{,}6)} = 0.0066\cdot10^{-3} \text{ or } 0.0066\%.$$

Thus, it is possible to calculate molar fractions of all macro components (column 4 in the Table below). In total for their share is necessary only 7.7·10^{-5} or 0.077‰ of all moles of the solution. Therefore, the molar fraction of moisture is very close to 1.

Molality describes the ratio of the number of component moles to the solvent mass expressed in kilograms. In the case of SO_4^{2-} it is equal to:

$$C_{m,SO_4} = \frac{C_{w,SO_4}}{M_{SO_4} \cdot (d - 0.001 \cdot m)} = \frac{0.0358}{96 \cdot (1.013 - 0.001 \cdot 0.1836)} = 0.37 \cdot 10^{-3} \text{ mole} \cdot \text{kg}^{-1} \, H_2O.$$

Similarly it is possible to calculate the molality of the remaining mineral components (column 5 in the Table below). As we see, values of molarities and molality are almost equal.

At last, the content of electrolytes is convenient to compare by the number of equivalent masses. For the determination of sulfate ion normality we will use this equation:

$$C_{eq,SO_4} = \frac{C_{w,SO_4} \cdot z_{SO_4}}{M_{SO_4}} = \frac{35.8 \cdot 2}{96} = 0.75 \text{ meq} \cdot \text{l}^{-1}.$$

According to this equation the sum of equivalents of all mineral components is equal to 5.42 meq $\cdot l^{-1}$. This value allows estimating the share of each component in formation of water electric charge. For sulfate ion this fraction is determined from the following equation:

$$\left[C_{SO_4}\right] = 100 \frac{C_{eq,SO_4}}{\sum_{1}^{n} Ceq, j} = 100 \frac{0.75}{5.42} = 13.75\%.$$

In this analysis the equivalent sum of all cations 2.71 meq$\cdot l^{-1}$ (50 meq -%) is equal to a similar sum of anions 2.71 meq $\cdot l^{-1}$ (50 meq -%). Therefore, analysis is performed correctly.

Component	mg$\cdot l^{-1}$	meq$\cdot l^{-1}$	Molar fraction, %	meq $\cdot kg^{-1}$	meq$\cdot l^{-1}$	meq -%
1	2	3	4	5	6	7
Na^+	28,5	1,24	0,022	1,22	1,24	22,87
Ca^{2+}	24,2	0,50	0,011	0,60	1,21	22,27
Mg^{2+}	3,2	0,13	0,0023	0,13	0,26	4,86
Cl^-	38,9	1,10	0,020	1,08	1,10	20,24
SO_4^{2-}	35,8	0,37	0,0066	0,37	0,75	13,75
HCO_3^-	53	0,87	0,015	0,86	0,87	16,02
Total	183,6	4,31	0,077		5,42	100,00

Sometimes water composition is expressed through concentrations of assumed dissolved salts. Such composition is called ***saline water composition***. This format of expressing the composition is used relatively infrequently, when the main object of study is not so much the solution

Table 1.6 Calculation sequence of ground water salt composition in meq·l^{-1}.

	CO_3^{2-}	HCO_3^-	SO_4^{2-}	Cl^-
Ca^{2+}	1	3	7	12
Mg^{2+}	2	4	9	11
Na^+	5	6	8	10

as the salt, which it contains and is capable of precipitating at evaporation. As a regular rule, such salt composition is calculated based on equivalent component composition of water and is shown in molar or weight units.

This composition is determined sequentially from less soluble salts to greater than soluble ones (Table 1.6). In connection with this first, is determined the content of least soluble carbonate salts in the order of Ca^{2+}, Mg^{2+} and Na^+, then sulfates, and last chlorides. Salt content of each anion is determined by its depletion.

If the salt composition is presented in M_lL_m format, then each of its moles needs l moles of cations and m moles of anions. For this reason the number of moles in any dissolved salt is possible to determine from the equation:

$$C_{M,salt} = \frac{C_{eq,M}}{l \cdot z_M} = \frac{Ceq,L}{m \cdot z_L}, \qquad (1.11)$$

$C_{M,salt}$ is the concentration of dissolved salt in mmole·l^{-1}. Content of the salts dissolved in water is convenient to determine by the concentration of anions or cations for which l *or* m is equal to 1. The mole number of dissolved salt in this case is equal to $C_{eq,i}/z_i$. For the conversion of molar salt concentrations to weight it is convenient to use the table of their molar masses (Table 1.7).

Example 1.2 It is required to determine the salt composition of brine in lake Tanatar located in Kulunda steppe if it has the following component composition, in meq·kg^{-1}:

CO_3^-	HCO_3^-	SO_4^{2-}	Cl^-	Ca^{2+}	Mg^{2+}	Na^+
6,980.0	930.9413	1,396.8	5,047.7	10.0	7.1	14,338.2

The least soluble salt $CaCO_3$ will precipitate first. 1 mole of this salt removes from solution 1 mole of Ca^2 and 1 mole of CO_3^-. As the content of Ca^{2+} is lower than CO_3^-, the content of this salt is equal to the content of Ca^{2+} in mg-equivalent divided by its charge, i.e., 10.0/ 2= 5 mmole·kg^{-1}. After the precipitation of $CaCO_3$ in the solution remains 6,980.0-10.0 meq·kg^{-1}

Table 1.7 Molecular masses of salts for main macro components.

Salt	Molar mass, g·mole^{-1}	Salt	Molar mass, g·mole^{-1}
$CaCO_3$	100.09	$CaSO_4$	136.14
$MgCO_3$	84.32	$MgSO_4$	120.365
Na_2CO_3	105.99	Na_2SO_4	142.04
$Ca(HCO_3)_2$	162.102	$CaCl_2$	110.986
$Mg(HCO_3)_2$	146.327	$MgCl_2$	95.211
$NaHCO_3$	84.001	NaCl	58.443

of CO_3^-, which is greater than the content of Mg^{2+}. For this reason the content of the second salt $MgCO_3$ is equal to 7.1/ 2, i.e., 3.55 mmole·kg^{-1}. The remaining CO_3^- is capable to form with Na^+ a third salt, i.e., soda Na_2CO_3. Soda content is convenient to determine from the content of CO_3^-, which is equal to 6,980.0 –10.0 –7.1=6,962.9 meq·kg^{-1}. Therefore, the soda content is equal to 6,962.9/2=3,481.45 mmole·*kg*$^{-1}$. After the soda is formed, only Na^+ remains in solution in the amount 14,338.2 - 6,962.9 = 7,375,4 meq·kg^{-1}. This remainder is equal to the sum of all remaining anions 930.9+ 1,396.8+5,047.7 = 7,375.4 meq·kg^{-1}. This means that HCO_3^-, SO_4^{2-} and Cl^- form salts only with Na^+ in the amount equal to 930.9 mmole·kg^{-1}, 1,396.8/2 mmole·kg^{-1} and 5,047.7 mmole·kg^{-1}, respectively (see the following table). If it is necessary to present the content of these salts in mass units, it is necessary multiply their content in molar format by the corresponding values of equivalent mass. The results of such conversion are also represented in the Table:

	Salt composition		
Salts	in mmole·kg^{-1}	in g·kg^{-1}	in % of mass
$CaCO_3$	5.0	0.5	0.06
$MgCO_3$	3.55	0.3	0.04
Na_2CO_3	3481.45	369.0	43.81
$NaHCO_3$	930.9	78.2	9.28
Na_2SO_4	698.4	99.2	11.78
NaCl	5047.6	294.9996	35.03
Bcero		842.2	100.0

1.3 Gas Components

Those ***Gas components*** of the water composition, which in pure form under normal atmospheric environment (atmospheric pressure and temperature

298 °K) are found only in gas state, are studied and used, as a regular rule, as a gas solution, and are called ***natural gas***.

Gas components are formed by elements of the upper right corner of the Mendeleyev's table, namely, H, C, N, O, S and the so-called inert or noble gases. Their molecules have covalent bond and do not have electric charge. Due to this many of these components are poorly soluble in water (no greater than 50 $cm^3 \cdot l^{-1}$). The exceptions are CO_2 and H_2S, which actively react with water and for this reason, are well soluble in it. Solubility of gas components depends not so much on temperatures as on pressure. At oversaturation they are released as bubbles, which float and are lost into the atmosphere. This feature in their behavior under normal conditions determines specifics of their testing and chemical analysis.

Analyzed gas components are convenient to separate into three groups. The first one includes inert gas components, which possess the highest chemical passivity and volatility. This group includes molecular nitrogen N_2 and noble (inert) gases He, Ne, Ar, etc. These components are extremely passive in chemical processes and possess the lowest solubility in water (9-34 $cm^3 \cdot l^{-1}$). They do not affect water quality.

Second group includes organic gas components, namely, methane (CH_4), ethane (C_2H_6), propane (C_3H_8) and two butane isomers (nC_4H_{10} and iC_4H_{10}). Their common features are similar genetic nature, similar chemical properties and also low solubility (20–45 $cm^3 \cdot l^{-1}$). These components are chemically very passive, present in surface and ground waters usually in minuscule amounts and, as a regular rule, do not present an ecological hazard.

The third group encompasses the remaining, very diverse, sometimes quite active gas components, which are rarely encountered together. Among them is usually define the content of O_2, CO_2 and H_2S, much rarer H_2 and CO.

Molecular oxygen O_2 does not react with H_2O (solubility 31.1 $cm^3 \cdot l^{-1}$) but it is a strong oxidant and easily converts to a fixed state. Its major source is the process of photosynthesis. For this reason the largest O_2 concentration is observed in the atmosphere and in surface and ground waters. Dissolved oxygen does not have hazardous physiological effects. On the contrary, its presence is necessary for life of any aerobic organism. O_2 content in the water of an average river of less than 2.0 $cm^3 \cdot l^{-1}$ may be considered as dangerous or deadly for fish. For the support of fish faunas in good conditions the dissolved O_2 concentration must be no lower than 4.0 $cm^3 \cdot l^{-1}$. The water oversaturated with dissolved O_2 also may be hazardous for fish. Besides, O_2 in water causes corrosion, which facilitates destruction of materials of engineering facilities.

Carbon dioxide CO_2 interacts with H_2O and forms carbonic acid, which dissociates into CO_3^{2-} and HCO_3^-. Due to this the CO_2 solubility in normal conditions reaches 878 $cm^3 \cdot l^{-1}$. Most CO_2 is a product of photosynthesis and one of sub-products of organic matter decomposition. However, a significant fraction of this CO_2 on the surface is absorbed by the vegetation in the process of the photosynthesis. For this reason surface and ground waters contain relatively little CO_2. Carbon dioxide in potable waters does not have physiologic value for humans and cattle but noticeably influences fish. 50% of the surface water, which support good fish fauna contains less than 0.8 $cm^3 \cdot l^{-1}$ CO_2 and 95% less than 2.5 $cm^3 \cdot l^{-1}$. The free CO_2 facilitates solution of carbonate minerals, and in particular, destruction of building materials containing lime, including concrete.

Hydrogen sulfide H_2S is an acid, which dissociates in water into HS^- and S^{2-}. Its solubility under normal conditions reaches 2,510 $cm^3 \cdot l^{-1}$. Hydrogen sulfide and sulfides form in ground waters as a result sulfate reduction. Their presence may be discovered in wastewaters of leather-dressing, paper, chemical and gas industries. Presence of sulfides, especially hydrogen sulfide, is undesirable not only because of their smell, but because hydrogen sulfide and its derivatives are a danger for many organisms. In potable waters content of hydrogen sulfide must not exceed 0.003 $mg \cdot l^{-1}$.

Determination of the content of gas components in ground waters also includes 3 stages: testing, chemical analysis and recalculation of its results. However, due to chemical passivity and high volatility of gas components these study methods are radically different from the techniques utilized with the mineral components.

1.3.1 Testing and Preparation

At ground water discharge on the surface, with testing and transportation of samples in laboratory conditions, most gas components usually are lost into the atmosphere. For their protection special testing methods are required, which include collection of natural gas samples.

Natural gas is a gas solution of most volatile components recovered from natural waters. Process of its recovery (extraction) is called ***degassing***. The volume of natural gas extracted from a unit volume of water is called ***gas-water ratio***. The natural gas should not be confused with the ***underground gas***, which is gas solution in conditions of a geologic medium.

The total process of ground water testing for dissolved gas components includes extraction of the natural gas, collection of its samples and determination of the gas-water ratio.

At the extraction of gas-saturated ground water, on the surface external pressure and temperature decline and part of gas components are released spontaneously. Such degassing, due only to a decrease in the external pressure and temperature is called ***spontaneous degassing***. Natural gas formed due only to the spontaneous degassing is called ***spontaneous natural gas***. Its volume and significance usually increase along with the depth of the studied water. The spontaneous degassing may be controlled and uncontrolled.

At uncontrolled degassing the natural gas is not tied in with specific volume of water. Such degassing occurs in the process of head water discharge onto the surface or at their spouting (natural flow) from the well. In this case the natural gas and degassed water come to the surface separately, and gas-water ratio is determined as the ratio of their stabilized rates. For the most complete extraction of the natural gas and measuring its gas-water ratio at uncontrolled degassing, special ***gas-water separators*** are often used. These separators provide for maximum water degassing at 1 atm. And enable separate determination of the water and natural gas rates. At the uncontrolled spontaneous degassing, the value of gas-water ratio is determined as the ratio of gas and waters rates.

For controlled degassing special ***depth samplers*** (***subsurface sampling devices***) are used (Figure 1.6), which enable collecting water samples at depth prior to the start of their spontaneous degassing. In the testing the sampler is run to the assigned depth, hermetically closed and recovered on the surface together with the water sample, preserving the internal formation pressure. Extraction of gas samples from the sampler and determination of its gas-water ratio is done using the gadget called ***adapter-connector*** (Figure 2.23). Gas-water ratio in this case is determined as ratio volume of extracted natural gas to the volume of the water in the sampler. Controlled degassing provides for the most reliable data on the gas composition of depth ground water.

In any spontaneous degassing some gas components remain in dissolved state. At high gas-saturation, when fraction of the spontaneous natural gas in total gas-saturation is close to 1, this residue is disregarded. However, with the decrease of gas-saturation the relative role of such remainder increases. In ground and surface waters spontaneous degassing, as a regular rule, is not present. At the gas-saturation below 0.1 $m^3 \cdot m^{-3}$ forced degassing is usually used.

Forced degassing is achieved either by lowering the external pressure much below the atmospheric – ***vacuum degassing***, or by heating the water

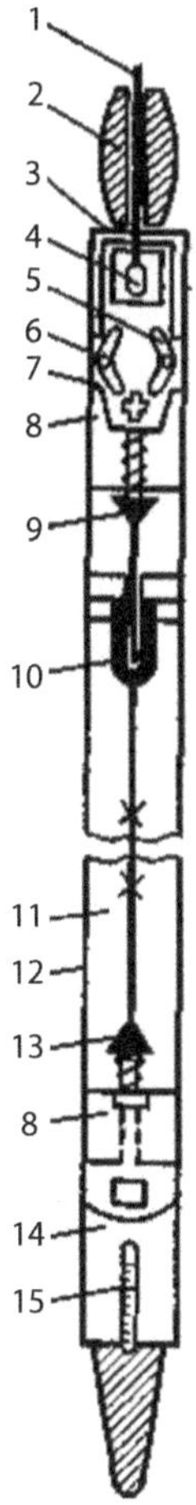

Figure 1.6 Conceptual schematics of downhole sampler PD-03.

The sampler with open valves 9 and 13 is run down the hole on a steel wire rope 1 (diameter 1.6—1.2 mm; rope holder 4). In the process the sampler gets the water through its hollow 12. At the testing depth the sampler is flushed (it is raised and downed by 10—15 m) and closed. Sampler valves PD-03 and PD-3M are being closed with the clock-work at the set time or with impact weight. In the latter case at the needed moment the load 2 is run down the rope. It pushes the impact device 3, which with the adapter-connector 8, turns about their axes, triggers 5 and 6 and closes the upper valve 9. At the same time the guide bar 7 is released from the lock 10, releases the motion rod 11 and closes valve 13. Housing 14 includes maximum thermometer 15.

Total mass of the sampler is 10 kg, length of the assembled device (together with thermometer) is 2.4 m, diameter 38 mm, operational capacity 0.85—0.80 l (V.N. Kortsenshtein, 1991).

to 100 °C – ***thermal degassing***. The first technique is technically greater than complex but greater than correct as it eliminates the influence of high temperatures on water-dissolved gas composition. Special degassers of various design are used for vacuum degassing. The device proposed by V.P. Savchenko is relatively simple and very popular (Figure 1.7). It is often called "Savchenko's flask". In the second case is used thermal degasser (Figure 1.8). High temperature may cause the decomposition of bicarbonates and pitch the estimate of CO_2 content and the gas-water ratio value. However, this drawback should not noticeably affect the hydrocarbon, N_2 and rare gases concentrations.

The gas-water ratio in the forced degassing is determined as the ratio of extracted natural gas volume to the volume of degassed water.

For safekeeping of natural gas samples are usually used glass vessels (bottles), greater than rarely vessels made of polymer material are used and very rarely, from rubber. As the testing is conducted in the air medium, the extraction of the natural gas and transferring it into bottles requires certain skill in order to eliminate or lower sample contamination with air. A natural gas sample is collected into a glass bottle by way of displacing by water. For this purpose the bottle preliminarily filled up with the studied water or brine is turned upside down within another vessel (bucket) filled with the same water. Dexterity should be used to avoid getting air bubbles into

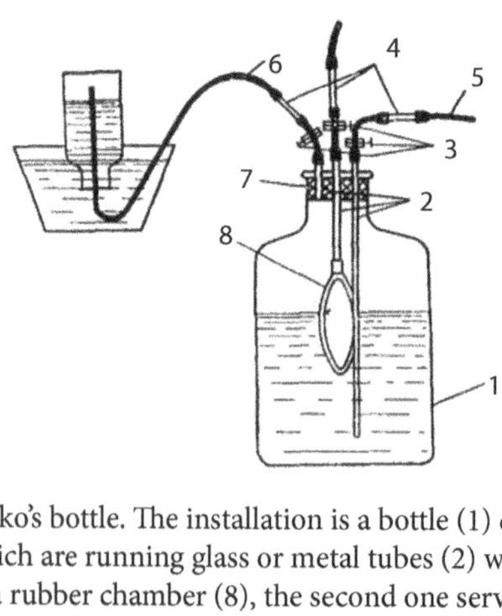

Figure 1.7 V.P. Savchenko's bottle. The installation is a bottle (1) closed by a rubber stopper (7), through which are running glass or metal tubes (2) with clamps (3). To one of the tubes is attached a rubber chamber (8), the second one serves for introducing the water through the hose (5) and is lowered almost to the bottom, and the third one is used for collection of the gas through hose (6).Using a pump, the air is removed from the empty bottle 1 and from the chamber 8. Then through the hose 5 the assigned volume of the studied water is introduced. The water is degassed under vacuum and the released natural gas is gathered in the upper part of the bottle. This gas then is displaced into the test bottle through hose 6 and inflates the rubber chamber 8.

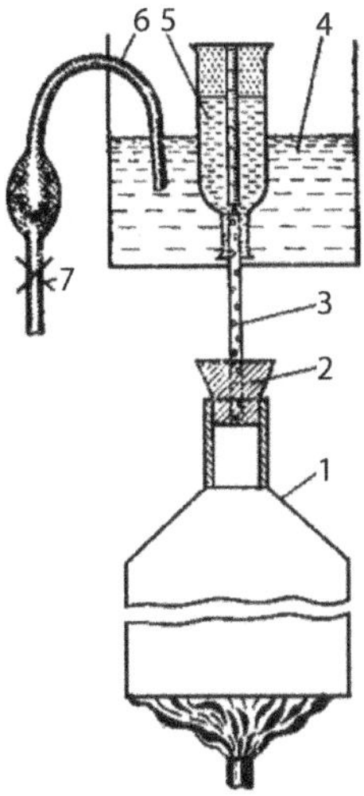

Figure 1.8 V.E. Vasserberg's thermal degasser is a hermetically closed vessel (1), which has the stopper (2) with a hole. Gas-conducting brass tube (3) is run through the hole. This tube runs through a cooler (4) and transfers the gas into the calibrated vessel for collecting the gas (5). The water in the cooler is changed through a tube with syphon (6) and clasp (7).

the bottle. Natural degassing gas is introduced into the bottleneck from the down up through a rubber hose (Figure 1.7). Gas floats up to the bottle bottom and gradually displaces the water. Usually a small water lock is left above the stopper. The bottle is corked with a rubber or plastic stopper, extracted from the water and stored upside down until needed.

The sum of gas-water ratios from the spontaneous and forced degassing gives ***the complete gas saturation*** of underground waters. However, at high gas content the second addend is usually disregarded. Value of gas-water ratio is usually expressed in $cm^3 \cdot l^{-1}$ or $m^3 \cdot m^{-3}$.

The measured volume of studied water is placed in hermetic vessel and heated to 100°C. The formed steam-gas mix is run through the cooler where the steam is condensed and the dry natural gas is transferred into the vessel for testing. (V.N. Kortsenshtein, 1991).

Thus, for the determination of gas components content, it is necessary, as a regular rule, to extract the natural gas and determine the gas-water ratio. Both are performed under field conditions and require special equipment, therefore, additional costs. For this reason uncontrolled spontaneous degassing without measuring the gas-water ratio is often chosen. In which case content of gas components in water is substituted by their content in the natural gas. This is an egregious and unacceptable error.

Table 1.8 Methods of sample preparation for gas component analysis.

Operations		**Components**
Conservation	To add: a) 1-3 ml $MnCl_2$ or $MnSO_4$ per 0.3 l; b) 1-3 ml alkali-iodine solution (NaOH with KI) per 0.3 l; c) 2 ml H_2SO_4 per 0.3 l.	O_2
	Add 100 ml of the saturated $Ba(OH)_2$ solution per 0.5 l	Total CO_2+HCO_3^-
	Add 2-3 g of $CaCO_3$ powder	Corrosive CO_2
	Add 100 ml of 4% $Cd(CH_3COO)_2$ or $Zn(CH_3COO)_2$ solution per 0.5 l	H_2S+HS^-
	Prevent contact with the air	He, H_2, Ne, O_2, Ar, N_2, CO_2
Concentrating	Spontaneous or forced degassing, collection of samples through displacing by water without catching air	He, Ne, Ar, N_2, O_2, CH_4, CO_2, H_2, H_2S, hydrocarbons, etc.

Table 1.9 Analysis techniques of ground water gas components.

Analyzed components	**Technique of analysis**
N_2, Ar, He, O_2, CO_2, CH_4, C_2H_6, C_3H_8, C_4H_{10}, etc.	Volumetric
	Chromatographic
	Mass-spectrometric
O_2	Iodimetric (Winkler or Alsterberg)
	Ionometric
CO_2	Monometric
	Titrating Na_2CO_3 or NaOH
H_2S	Iodimetric

In the absence of spontaneous degassing, the O_2, CO_2 and H_2S content is possible to determine directly in the natural water by titration. These analyses are performed either directly in the field or in a laboratory with the preliminary conservation (Table 1.9).

1.3.2 Analysis of the Natural Gas Composition

The second important step in the determination of the ground water gas composition is analysis. In most cases the composition of the natural gas extracted from water or the content of O_2 , CO_2 and H_2S are analyzed directly in the water.

Difficulties in natural gas analysis are caused by high volatility and chemical passivity of its components but also by the risk of sample contamination by the air. For this reason natural gas analyses are performed mostly by physical methods, often in conditions of a high vacuum and strict isolation from the atmosphere.

Natural gas analysis is conducted with volumetric, chromatographic, spectral, mass-spectral and other methods (Table 1.10). Initially the leading technique was the classical volumetric (volume, absorption), which is based on a selective and sequential removal from the natural gas of its components by solid or liquid chemicals. Due to this the gas mix volume decreases. Measuring the changes in the volume or pressure allows calculation of the content of the removed gas component. This technique is very complex, painstaking, and requires high vacuum but allows separation of gas components in pure form for their subsequent isotope analysis.

Currently most gas samples are analyzed using the technique of gas chromatography. This technique is based on different sorption properties of analyzed components. A greater than detailed description of the analysis technique is provided below.

The natural gas composition analysis results are usually given in volume percentages. As one mole of any gas component under normal conditions occupies one and the same volume (22.4 l), the volume percentages are equal to molar percentages. The sum of all gas component concentrations in the natural gas must be equal to 100%. However, the nature and number of components analyzed in the natural gas are not always equal. As a regular rule, in natural gases oxygen – O_2, nitrogen – N_2 , carbon dioxide – CO_2, hydrogen sulfide – H_2S and lower alkanes (saturated hydrocarbons), namely, methane – CH_4, ethane – C_2H_6, propane - C_3H_8 and isomers of butane – C_4H_{10} are determined. Much greater than rare are determined hydrogen – H_2, carbon monoxide – CO, heavier hydrocarbons, helium and argon.

At last, in the absence of spontaneous degassing chemically active O_2, CO_2, H_2S and Rn (radioactivity) may be determined directly in the solution.

Molecular oxygen at low gas-saturation is determined by Winkler's test (or Alsterberg's test) based on iodimetry. For this purpose immediately after the collection the equivalent content of free iodine is substituted for samples of the dissolved oxygen; iodine is greater than stable in solution than oxygen. Then, by titring solution of the sodium thiosulfate $Na_2S_2O_3$ the content of this iodine is determined and O_2 volume is calculated. This analysis is hindered by the presence of unrelated oxidants or deoxidants (iron, nitrite, organic matter, etc.), which understate or overstate derived results. In order to eliminate their influence, different modifications of Winkler's technique exist. Usually the reviewed technique is used with O_2 concentrations no lower than 0.2-0.3 mg·l^{-1} but in the absence of spontaneous degassing. In cases of not very polluted waters its results are reproducible to ±0.01 mg. However, at concentrations below 0.1 mg·l^{-1} this error may reach ±50 %.

In recent times are actively introduced physical analysis methods of the free O_2 content metering in natural and waste waters with oximeter. This technique is based on the correlation of diffusion current amperage at O_2 reduction on the electrode vs. O_2 concentration in the solution. Oxygen penetrates to the electrode through a special polymer water-impermeable membrane in the amounts in proportion to its concentrations in water. Portable oximeters AzhA-101.1M, ANION-7041, "Super Dist 5" and others (Figure 1.9) enable the metering of O_2 directly in field conditions over a wide range of concentrations and with high reliability.

Free carbon dioxide at low gas-saturation is possible to determine directly in the water sample. According to GOST 23268.2—91 this content in mineral potable, balneological, curative-table and table waters should be determined by the monometric or titrimetric technique.

Monometric technique is used for a rough estimate of the CO_2 content at presence of low spontaneous degassing. It involves measurements of

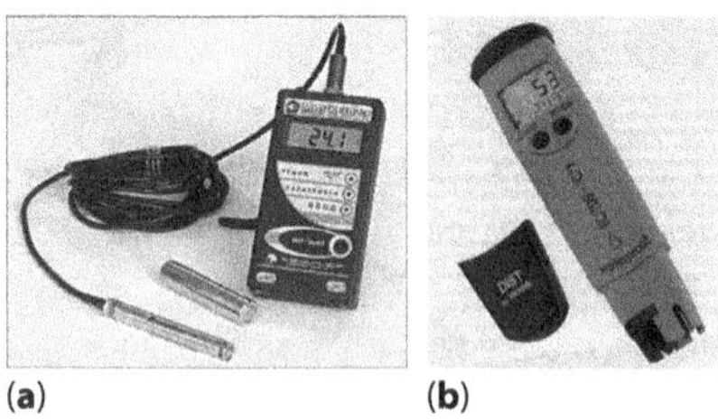

(a) (b)

Figure 1.9 Water testers for metering temperatures, electric conductivity and oxygen concentrations. A – oximeter MARK-302T. Oxygen metering time not to exceed 60 s. Range O_2, mg·l^{-1}: 0–20, measurement resolution, O_2: 0.001, measurement error, O_2: ± 3 μg·l^{-1}+4% of measured values. B. Water tester Super Dist 5: Range O_2, mg·l^{-1}: up to 1999, measurement resolution, O_2: 1 mg/l, measurement error, O_2: ±2% of measured values.

pressure changes in hermetic vessel after a water sample is placed in it and all spontaneous gas is allowed to escape. Presence of other gas components facilitates overestimating of the CO_2 content.

Titrimetric technique is based on the conversion of free CO_2 in hydrocarbonate ion by an alkali (Na_2CO_3 or NaOH) and determination of the HCO_3^- content by ittering by an acid to pH 8.3-8.5. Weight content of free CO_2 is calculated as follows:

$$C_{\mathrm{w,CO_2}} = 0.72\left(C^*_{w,\mathrm{HCO_3^-}} - C_{\mathrm{w,HCO_3^-}}\right) \tag{1.12}$$

where $C_{\mathrm{HCO_3^-}}$ and $C^*_{\mathrm{HCO_3^-}}$ are weight contents of HCO_3^- without the addition of alkali and with its addition. Reliability of this technique is hindered by the presence of other weak acids, first of all organic.

Free carbon dioxide may be equilibrium or excessive, i.e., corrosive. At pH 6 – 9 and salinity less than 0.5 g/l, free equilibrium CO_2 is possible to estimate by calculation from the HCO_3^- content:

$$C_{\mathrm{o,CO_2}} \approx K \cdot 10^{(10-\mathrm{pH})} C_{M,\mathrm{HCO_3}}. \tag{1.13}$$

where K is a factor whose value decreases from 1.22 at 5°C to 1.06 at 10°C to 0.88 at 20°C and до 0 $ml \cdot mg^{-1}$ at 30°C.

Example 1.3 In the water of a spring running from Ordovician limestones near Skvoritsa village (Leningrad Province), at pH 7.2 and temperature 7°C was discovered 420 $mg \cdot l^{-1}$ of HCO_3^-. At 7°C, K value *is* equal approximately to 1.16 $ml \cdot mg^{-1}$. Because of this the volume concentration of equilibrium dissolved CO_2 in the spring water is equal to:

$$C_{\mathrm{o,CO}} \approx 1.16 \cdot 10^{(6-7.2)} 420 = 30.7\,\mathrm{ml} \cdot \mathrm{l}^{-1}$$

Hydrogen sulfide is usually determined by iodine-metering technique in combination with sulfides but then H_2S and HS^- fractions are calculated. This technique is based on the reaction of oxidizing H_2S by iodine. The presence of heavy metals facilitates underestimation, and the presence of oxygen and other oxidants – overestimation, of analysis results. At sulfide content below 0.05 $mg \cdot l^{-1}$, reproducibility of the results declines to ±50%. With the increase of sulfide concentrations, their reliability increases, and at 5 $mg \cdot l^{-1}$ the results are reproducible to ±0.05 mg. For this reason this technique is recommended at the sulfide content in excess of 0.5 $mg \cdot l^{-1}$.

Contents of O_2, CO_2 and H_2S analyzed directly in the solution, as a regular rule, are expressed in weight units.

Radon is a radiogenic and radioactive gas component of a certain ecological danger. Radon with its affiliated products of radioactive decay is responsible for about 3/4 of the annual individual effective equivalent irradiation dose inflicted on the population from Earth radiation sources, and approximately for half of this dose from all natural radiation sources. For this reason, despite this gas being present in water in negligibly small amounts, it was in recent times a subject of significant attention. Radon content is determined based on its radioactivity value. Radon content is expressed in units of specific radioactivity, namely, in becquerels $Bq{\cdot}m^{-3}$ or curies $pCi{\cdot}l^{-1}$.

1.3.3 Conversions of Gas Analysis Results

Natural gas analysis results provide the content of individual components in fractions or percentages of its volume. Total content of these components must be equal to 1 or 100%. Noticeable deviation from this value indicates an error or an incomplete analysis.

If analysis results include oxygen, it is necessary to determine its nature. This oxygen may belong with the studied natural water, in which case it has hydrochemical interest. But it also could have entered the water in testing, safekeeping of the samples or even while transferring it into the analysis device. There are no reliable criteria for the identification of these O_2 sources, and for shallow ground waters this problem is irresolvable. Only in deep waters, where oxygen does not have analytically noticeable concentrations, its presence uniquely indicates sample contamination with air. In such cases analysis results are recalculated by way of excluding foreign atmospheric components.

Exclusion of air from analytical composition includes two operations: 1. Direct removal of its components and 2. Conversion of natural gas composition after this removal.

The first operation removes the entire O_2 and associated N_2 and Ar. The presence of the remaining air components is usually disregarded. Concentration ratios of the components in the air composition are permanent: $N_2/O_2 = 3.727$, $Ar/O_2 = 0.0444$. For this reason, if oxygen is completely eliminated, the content of molecular nitrogen and argon decrease, respectively, by values $3.727\ C_{nat,O2}$ and $0.0444\ C_{nat,O2}$, where $C_{nat,O2}$ is content of molecular oxygen in the natural gas. Due to this the volume of gas sample should be decreased by the total of $C_{nat,O2} + 3.727\ C_{nat,O2} + 0.0444\ C_{nat,O2} = 4.77\ C_{nat,O2}\%$.

In order to determine the natural gas composition prior to contamination, it is necessary to increase all concentrations remaining after air exclusion by as many times as the sample volume decreased, i.e., to multiply it by the factor k, which is equal to:

$$k = \frac{\sum_{j-1}^{J} C_{\text{nat},j}}{\sum_{j-1}^{J} C_{\text{nat},j} - 4{,}77\,C_{\text{nat,O}_2}} \approx \frac{1}{1-0.0477C_{\text{nat,O}_2}} \tag{1.14}$$

Here $C_{\text{nat,j}}$ is the content of any gas component i in the natural gas sample. It follows from this that in the updated analysis oxygen O_2 will be absent and the nitrogen and argon content will be determined from the following equations:

$$\begin{aligned} C^*_{\text{nat,N}_2} &= k\cdot\left(C_{\text{nat,N}_2} - 3.727\cdot C_{\text{nat,O}_2}\right) \text{ and} \\ C^*_{\text{nat,Ar}} &= k\cdot\left(C_{\text{nat,Ar}} - 0.0444\cdot C_{\text{nat,O}_2}\right). \end{aligned} \tag{1.15}$$

The concentrations of the remaining components will then be equal to:

$$C^*_{\text{nat},j} = k.C_{\text{nat},j}. \tag{1.16}$$

Example 1.4 At spontaneous degassing of the underground water from a great depth the natural gas was released. The following composition of its sample was derived in gas laboratory: CH_4 – 80.95%, C_2H_6 – 1.9%, N_2 – 15.15%, O_2 – 1.0%, Ar – 0.52% and He – 0.48%. Oxygen in underground water at the testing depth must be absent. Its presence indicates the sample contamination by air. In this case the conversion factor is:

$$k = \frac{1}{1-0.0477\cdot 1} = 1.05.$$

And the updated natural gas composition is: CH_4 – 1.05·80.95=85%, C_2H_6 –1.05·1.9=2.0%, N_2 – 1.05·(15.15–1·3.727)=12.0%, O_2 – 0%, Ar – 1.05·(0.52 –1·0.0444)=0.50% and He – 1.05·0.48=0.5%.

Content of the components in natural gas describes relations of gas components between themselves, not relative to the solvent, i.e., H_2O. For the determination of the gas component contents in water it is necessary, beside the natural gas composition, to have the gas-water ratio value. Then the volume content of gas components in the underground water will be equal to:

$$C_{o,i} = 0.01\left(C_{o,i,spont.}Q_{spont.} + C_{o,i,nat.}Q_{nat.}\right) \quad (1.17)$$

where $C_{o,i}$ is volume content of the gas component i in water, $C_{o,i,spont.}$ And $C_{o,i,nat.}$ are concentrations of the same component in natural gas of spontaneous and forced degassing in volume percentages; $Q_{spont.}$ And Q_{nat} are values of gas-water ratios of spontaneous and forced degassing. Gas-water ratios of forced degassing are low, and at high gas-saturation they are disregarded. ***Volume concentration*** of gas components is usually measured in the same units as gas-water ratio, i.e., in $ml \cdot l^{-1}$ or $m^3 \cdot m^{-3}$, and greater than rarely in $ml \cdot kg^{-1}$ or $m^3 \cdot ton^{-1}$. Content of O_2, H_2S and CO_2 determined directly in water is often expressed in $mg \cdot l^{-1}$ or $mg \cdot kg^{-1}$. Volume and weight concentrations of gas components are found from this equation:

$$C_{o,i} = 22.414\frac{C_{w,i}}{M_i}, \quad (1.18)$$

where $C_{o,i}$ is the volume content of a component i, in $ml \cdot l^{-1}$, $C_{w,i}$ is weight content of the same component i, in $mg \cdot l^{-1}$.

The numerical content of gas components is expressed in moles. As the mole of any gas component under normal conditions occupies the volume of 2.2414 10^4 cm^3, the number of moles is determined so:

$$C_{M,i} = \frac{C_{o,i}}{2.2414 \cdot 10^4} = 4.4615 \cdot 10^{-5} C_{o,i}. \quad (1.19)$$

For the study of interaction between the gas components in different media (water and oil, water and underground gas) it is often convenient to express their concentrations in molar fractions:

$$C_i = \frac{C_{M,i}}{\sum_{j=1}^{n} C_{M,j}} = \frac{C_{M,i}}{C_{M,H_2O} + \sum_{j=2}^{n-1} C_{M,j}} \approx \frac{C_{M,i}}{C_{M,H_2O}} \approx 18\frac{C_{M,i}}{\gamma_w - m}$$

$$= \frac{18 \cdot C_{o,i}}{22.414 \cdot \left(10^6 \gamma_w - m\right)} = 0.8039\frac{C_{o,i}}{\left(10^6 \gamma_w - m\right)}, \quad (1.20)$$

where C_i is concentration of the component i in water, in molar fractions; $C_{o,i}$ is content of the same component in $ml \cdot l^{-1}$; γ_w is water specific gravity, $g \cdot ml^{-1}$, m is water salinity, $mg \cdot l^{-1}$, 18.016 g $mole^{-1}$is the molecular mass of H_2O.

Correlation between gas components' solubility in water and pressure values predetermines one greater than parameter associated with their concentration, namely, ***partial pressure***. Strictly speaking, partial pressure is pressure of a separate gas component in the gas solution. As ***partial pressure*** of any gas component in water is accepted its partial pressure in gas phase in equilibrium with this water. Partial pressure is measured in atmospheres, bars or Pascals. The parity of gas components' partial pressures in different media does not mean parity of its concentrations. For the determination of gas components' partial pressures, ***solubility factors*** are usually utilized. These factors show the correlation between partial pressure of a component *i* and its concentration. In real life for the calculation of partial pressures in water, ***Bunsen's solubility factor*** B_i is commonly used, which is volume of component *i* per unit volume of water (in $cm^3 \cdot l^{-1} \cdot atm^{-1}$) at its partial pressure of 0.10 Mpa. Under low pressure conditions partial pressure is possible to determine, so:

$$p_i = \frac{C_{o,i}}{B_i}, \tag{1.21}$$

where p_i is partial pressure of the component *i* in water.

Thus, the presence of gas components in ground water may be shown either through their volume and molar concentrations or through their partial pressures. The way selected to express ground water gas composition depends on the task.

Example 1.5 Based on the results of the natural gas analysis in the previous Example, it is necessary to determine the gas composition of underground waters, provided the gas-water ratio is equal to 600 $ml \cdot l^{-1}$. The water salinity is 118 $г \cdot l^{-1}$. The water density is 1.08 $г \cdot ml^{-1}$, and temperature is +40° ***C***.

At such high gas-saturation, the contents of gas components remaining in water after spontaneous degassing may be disregarded. Then the volume content is determined from the following equation:

$$C_{o,i} = 0{,}01 \cdot C_{i,nat}\, Q.$$

Thus, we have the following contents of gas components in water: CH_4 – 0.01·85·600=510 $ml \cdot l^{-1}$, C_2H_6 –12 $ml \cdot l^{-1}$, N_2 –72 $ml \cdot l^{-1}$, Ar –3 $ml \cdot l^{-1}$ and He – 3 $ml \cdot l^{-1}$.

In order to determine the molar content it is sufficient to divide the volume content by the volume of one mole, i.e., by 22414 $ml \cdot mole^{-1}$. And we have: CH_4 – 510/22414=0.022754 $mole \cdot l^{-1}$, C_2H_6 – 0.000535 $mole \cdot l^{-1}$, N_2 – 0.003212 $mole \cdot l^{-1}$, Ar –0.000134 $mole \cdot l^{-1}$ and He – 0.000134 $mole \cdot l^{-1}$.

And at last, for calculating the molar fractions of gas components in underground water we will use this equation:

$$C_i = 0.8039\frac{C_{o,i}}{\left(10^6 d - m\right)}.$$

Within the framework of the problems being solved, water salinity and density are constant. For this reason it is convenient to express the constant values as one constant, $0.8039/(10^6 d-m) = 7.7310^{-4}$. And from this the molar fractions of gas components in underground water are: CH_4 $-7.7310^{-4}\cdot 510\cdot 10^{-3} = 0.426\cdot 10^{-3}$, C_2H_6 – $0.01\cdot 10^{-3}$, N_2 $-0.06\cdot 10^{-3}$, Ar $-0.0025\cdot 10^{-3}$ and He – $0.0025\cdot 10^{-3}$.

1.4 Organic Components

As ***organic components*** in this case are treated all organic compounds dissolved in natural water except gaseous hydrocarbons. These are numerous, complex, often volatile compounds of carbon and hydrogen in ground water. As opposed to mineral organic components, in the case of oversaturation they are released from solution in the form of various hydrophobic liquids, which have density, as a regular rule, less than $1 g\cdot ml^{-1}$ and float to the surface. They differ from gas components in that in pure form under normal circumstances they may exist as a liquid and gas simultaneously. Organic components recovered from water form in combination complex solutions called ***organic matter***.

Elemental composition of organic matter dissolved in water includes carbon (44.5–87.5%), hydrogen (5.5–13.1%), oxygen (9.1–49.3%), nitrogen (1.1–15.0%) and sulfur (less than 2.4%). An important element of organic matter is also phosphorus. These elements are constant components of living organisms and are called ***biogenic elements*** .

Component composition of organic matter is very complex as even at low content it is capable of containing many different compounds. Most of it possesses covalent bond, chemically is quite passive and poorly dissolves in water. The type of interaction between organic compounds and water, as a regular rule, is determined by the presence in their composition of functional groups, especially hydroxyl (-OH) or carboxyl (-COOH). These groups lose hydrogen relatively easily. Due to this the compounds acquire properties of weak acids soluble in water. Examples are alcohols or carboxyl acids. Besides, even minute changes in structure of an organic substance, such as changing place by the Cl atom on the benzene ring, may noticeably affect its solubility. It is convenient to subdivide organic components in ground waters into three groups: nonpolar, ion-forming (amphiphilic) and complex compounds.

Nonpolar ones are represented mostly by hydrocarbons. All these compounds have covalent bond, do not have charge and are chemically passive. They are poorly soluble in water, possess high resistance to acids, alkali and oxidants. Typical for them are substitution reactions. In particular, halogens and nitrogen are capable of displacing hydrogen and forming, respectively, halogen derivative hydrocarbons and amines. With growing size of molecules and their molecular mass the solubility noticeably decreases. Due to this, even at relatively low amounts, the nonpolar organic components tend to segregate from water solution and take the form of independent liquid (oil) or gas (underground gas) phase. The hydrocarbons dissolved in water are dominated by saturated alkanes (methane, paraffin), cyclo-alkanes and aromatics.

The saturated alkanes have formula C_nH_{2n+2}. The lightest among them: methane, ethane, propane and butanes are gas components. Among organic matter are studied pentane and heavier homologs of methane, which under normal conditions are liquids or even solid matter. Unsaturated noncyclical hydrocarbons (alkenes and alkines) are chemically highly active and unstable under natural conditions. This is a reason for their almost total absence in ground waters. The content and solubility of alkanes decreases with the increase of the number of carbon atoms.

Cyclo-alkanes (naphthenes) are saturated cyclic hydrocarbons (C_nH_{2n}). They take second place in the abundance. They are even less volatile and practically insoluble in water.

Aromatic hydrocarbons are molecules with one or several benzene rings (C_nH_{2n-6}).

All hydrocarbons have capacity to replace hydrogen with halogens, in particular chlorine, forming halogen derivatives, and with nitro-group $-NO_2$, forming nitro-compounds. They are also capable of replacing hydrogen in ammonia forming amines.

Ion-forming (amphiphilic) organic compounds contain functional groups capable of losing hydrogen (-OH, $-NH_3$, etc.). Under H^+ deficiency they form charged anions. Due to this their solubility in water noticeably increases. In this group should be identified, first of all, alcohols, carboxylic acids and phenols.

Alcohols are hydrocarbons with one or several –OH groups. They have very high solubility in water, which decreases with the increase in their molecular mass. But they easily oxidize and are encountered in ground waters only in minuscule amounts. In the food chain they are easily destroyed and are not toxic. Elevated concentrations of some alcohols (methanol, glycerin, ethylene-glycol and butanol) are the result of contamination.

The most stable and common in this group of compounds are carboxylic acids, the most oxidized forms of hydrocarbons. They include organic compounds with functional group OOH. Examples are acids: formic HCOOH, acetic CH_3COOH, propionic CH_3CH_2COOH, butyric $CH_3(CH_2)_2COOH$, benzonic C_6H_5COOH, palmic $C_{16}H_{32}O_2$ and searic $C_{18}H_{36}O_2$. Their share is 5–8% of surface water organic carbon. In deep formation waters their content may reach 5 $g{\cdot}l^{-1}$. Their formation is associated with the decomposition and oxidizing of organic matter. Technogenic sources of these acids are garbage dumps, food industry refuse and phenoxyacetic pesticides, such as 2,4-D (2,4-dichlorophenoxyacetic acid), trichlorophenoxypropionic acid (2,4,5-TP or Silvex), etc.

Phenols are compounds which include the benzene ring with hydroxyl groups. They are represented in ground waters by phenol, cresols and syringic, vaniliac and *p*-hydroxybenzene acids. In non-polluted ground waters their content does not exceed 1 $\mu g{\cdot}l^{-1}$. Noticeable amount of phenols (phenol, cresol, pyrocatechol and naphtols) come from industrial waste or due to destruction of biocides (dinitrocresol, 2,4-dinitrophenol).

Complex organic compounds are the most cumbersome formations, which are difficult to systematize in terms of functional groups. It is possible to include among them chlorophyll, xantophyll, enzymes, asphalt-resinous matter and also humic acids. Natural pigments chlorophyll and xanthophylls usually have low solubility and represent a tiny portion of the composition of ground water as do enzymes. Much higher concentrations may have technogenic tanning matter (for instance, tannin $C_{14}H_{10}O_9$), colorants, luminophores, etc. Asphalt-resinous matter are macro-molecular compounds which include N, S, O and metals (V, Ni, etc.). They are poorly soluble in water and are encountered in minuscule amounts as an indication of the presence of oil or of contamination by petroleum products.

Organic matter of humic origin has special significance as the decomposition products of ***humus*** of plant remains. If such substance is treated with a strong alkali (NaOH), part of it, mostly complex organic acids, comes into solution. The totality of this organic matter is called ***humic acids***. They are conventionally subdivided into humic and fulvic acids according to their solubility in water of different acidity. The part of humic acids, which precipitates at acidification of the water, is attributed to ***humic acids***. The part of the dissolved organic matter, which remained in the solution at acidification, is called ***fulvic acids***. They are noticeably soluble in ground waters under normal conditions.

These acids include a number of polymer organic compounds with large molecular mass. Their acidic properties are due to the presence of numerous functional groups (mostly carboxyl and phenol). In the water these compounds often present in form of suspended (or colloid-dispersed) particles with well-expressed sorption properties and form amorphous dark-brown mass. The content of the fulvic acid is usually almost 10 times that of the humic acid.

Beside the aforementioned higher compounds in the composition of dissolved organic matter, it is possible to encounter esters, lipids, ketones, aldehydes, carbohydrates, etc. but they are all, as a regular rule, unstable, present in minuscule amounts, and nontoxic.

Hydrobiologists subdivide dissolved organic matter into three groups: acidic, base and neutral. The acidic one includes humic, carboxyl, naphthene acids, polyphenols, phenols and other organic acids. The base one includes cation-like proteins, amines, amino acids and complex compounds. Neutral organic matter includes sugars, polysaccharides, carbohydrates, etc.

Even at negligible content (fractions of a $\mu g \cdot l^{-1}$) some organic compounds may be life-threatening. Most toxic are halogen-derivative hydrocarbons and organometallic compounds, which enter ground waters with technical solvents (dichloromethane, chlorophorm, carbon tetrachloride, 1,1,1-trichlorethane, trichloroethylene, etc.), pesticides (Aldrin, Dieldrin, etc.) and other industrial waste (chlor-vynil, methyl chloride, iodomethane, mercur-ethyl, copper phtalocyanine, etc.). A special group is ***surface-active matter*** (***surfactants***). They are capable of adsorbing at the boundary between water and other matter. Surfactants comprise hydrocarbon group CH_2 (4 to 20 of them) and polar radical at the end ($-OH$, $-COOH$, $-NH_2$, $-SO_3H$, etc.). There are ion-genic surfactants, i.e., dissociating into ions (anion-genic and cation-genic) and non-ion-genic (amphoteric). As a regular rule, anion-genic surfactants dominate.

Many organic compounds are very toxic and their content in drinking water is strictly controlled. In Russia SanPiN 2.1.4.1074-01 (2002) records almost 670 dangerous organic compounds, which may be present in drinking water and may be identified by modern analytical methods. Out of those, 14 are extremely dangerous and around 240 are highly dangerous for human life. USA's EPA includes in a similar list of ***priority pollutant organic compounds*** 126 components. In view of this, the analysis of micro-concentrations of organic compounds in ground water in conditions of increasing technogenic load becomes of ever greater importance.

1.4.1 Testing and Preparation

Organic matter are present in ground waters, as a regular rule, in very small amounts and are poorly preserved in waters samples. Some of them rapidly volatilize together with gas components. Some others are easily oxidized and decomposed by microorganisms. In connection with these, the methods of collection and preparation of water samples for organic analysis depend to a significant extent on the properties, concentrations and preservation of the analytes.

If the duration of sample safekeeping does not exceed several hours (less than 24), it is sufficient to eliminate the contact with air and keep the sample at a temperature of 2-5°C in a dark place. If active chlorine is present, for its removal is added 20 mg of thiosulphate sodium $Na_2S_2O_3$ per 1 liter of sample.

If there is a need to safekeep a water sample for several days it, as a regular rule, is conserved. For this purpose either disinfectant matter (chloroform, timole, formaldehyde, etc.) is used or the sample is acidified to pH 2 and held at a temperature of 2-5°C, or even frozen at minus 10-20°C. For acidification is used HCl, H_2SO_4 or HNO_3. HNO_3 is acting both as acid and as oxidant. However, in samples with high alcohol content and in aromatic hydrocarbons, it is capable to interact with the resulting formation of explosive compounds like nitroglycerine or trinitrotoluene (trotyl).

The volume of a sample depends on the content and nature of organic matter and also on analysis technique. Usually at high organic content this volume is between 25 and 1,000 ml. However, a high amount of organic matter is present in ground waters in very low concentrations and is easily lost, which requires its concentration or division. Methods of these operations depend on volatility and solubility of organic components in water. Usually organic components are subdivided into volatile (VOL), semivolatile (SVOL) and nonvolatile (NVOL).

As nonvolatile (NVOL) are considered high-molecular components, which practically do not evaporate under normal conditions, or are indefinitely soluble in water. Due to this they actually are not lost at water degassing and at collection and safekeeping of their water samples under normal conditions. This group includes low-molecular polar compounds, such as amino-, oxy-, phenol- and poly-carboxylic acids, polyatomic phenols, surfactants, pesticides, etc. In their testing and conservation it is sufficient to eliminate oxidizing, biochemical processes and contamination by foreign matter. If conserved samples are nonvolatile, soluble, organic, collected in a glass, well-washed vessel, and cooled to 2-5°C, it is possible to hold them in darkness for several days (Table 1.10). If these organic analytes have very low concentrations, they are being concentrated. The technique of separation and concentration can have a very complex character for complex studies of organic matter.

Table 1.10 Technique of sample preparation for analysis of organic components.

Operation of preparation		Target organic components
Conservation	Filtration (0.45 μm)	Acetone, formaldehyde, chlorophyll
	Acidification to pH < 2	TOC, organic nitrogen , monocyclic arenes, crude oils and petroleum products, organic acids, lipids and oils (lubricants), phenolss, halogen-containing, chlor-phenolss, pesticides, anion-genic surfactants
	Alkilification to pH > 11	Phenolss, chlorphenolss
	Suppression of microorganisms	surfactant, crude oils and petroleum products
	Antisorbent (alkilic esther)	Surfactant
	Exclusion of contact with air	TOC, acetone, formaldehyde, monocyclic arenes (benzene), polycyclic aromatic hydrocarbons (benzopi-ren), crude oils and petroleum products, naphtene acids, halogen-containing, chlorphenolss, surfactants.
	Cooling to 2-5°C	TOC, organic nitrogen, alkanes, acetone, monocyclic arenes (benzene), polycyclic aromatic hydrocarbons (benzopiren), crude oils and petroleum products, naphtene acids, lipids, oils (lubricants), phenolss, chlorphenolss, haloethers, surfactants, pesticides, lignins and tannins
	Freezing to minus 20 °C	TOC, organic nitrogen, phenolss, chlorophyll
	To be kept in the dark	organic nitrogen, acetone, formaldehyde, monocyclic arenes (benzene), polycyclic aromatic hydrocarbons, phenolss, organic acids, halogen-containing, pesticides, chlorophyll
Concentrating	Gas extraction	Monocyclic arenes (benzene)
	Liquid extraction	TOC, alkanes, formaldehyde, monocyclic arenes (benzene), humic acids, polycyclic aromatic hydrocarbons (benzopiren), crude oils and petroleum products, phthalates, naphtene acids, humic acids, phenolss, halogen-containing, chlorphenolss, pesticides, lipids and oils (lubricants)
	Solid-phase extraction	Polycyclic aromatic hydrocarbons, humic acids, phenols
	Freezing-out	Phenolss, humic acids
	Distillation and rectification	Acetone, organic acids, humic acids, phenolss, amines, alcohols

Compounds are called semivolatile (SVOL) when they have boiling temperatures higher than water and noticeably evaporate under normal conditions. Their content in the process of water degassing changes insignificantly but noticeably decreases at long safekeeping of samples. To the category of semivolatiles belong organic acids, phenols, polycyclic aromatic hydrocarbons, etc. In collection and safekeeping of water samples for SVOL analysis it is important to observe hermeticity, i.e., isolation of the water of atmosphere. For this purpose vessels with samples are being completely filled, without any remainder of air, hermetically corked and held at temperature 2-5°C in a dark place.

The volatiles (VOL) include low-molecular compounds with high boiling temperature and low solubility in water. They comprise low-molecular hydrocarbons, including their halogen-derivatives and nitrogen-containing compounds, monoatomic phenols, low-molecular alcohols, carboxylic acids, ketones and aldehydes. These organic matter are often lost even before testing due to spontaneous water degassing. At large gas-water ratios, part of volatile organic components turns out with natural gases and part, in degassed water. For this reason the content of such organic components should be determined as the sum of their amounts in the natural gas and in the water. Samples of the natural gas for analysis of VOL are held in glass vessels, stopper down, with minimum seal (over the stopper) of tested waters, in darkness at temperature 2-5°C. The samples of water with volatile organic components also should be collected in glass (better, boronsilicate glass) well washed vessels and held cooled to 2-5°C. Nevertheless, it is not recommended to keep them in this form for greater than 1 day. For a longer safekeeping (over 1 month) the organic analyte should be either concentrated or extracted.

Concentration is the operation increasing the analyte content in waters whereas separation is increasing its content by way of extraction from solution. Both operations are performed either directly at the time of testing or before the analysis. Their major objective is to increase concentration, purity and preservation of studied compounds.

Quantitative indication of effectiveness of these processes is ***concentration factor*** or ***division factor***:

$$K_i = \frac{g_{i,\mathrm{end}}}{g_{j,\mathrm{end}}} \bigg/ \frac{g_{i,\mathrm{beg}}}{g_{j,\mathrm{beg}}}, \tag{1.22}$$

where $g_{i,beg}$ and $g_{j,beg}$ are mass component i and solvent j before the operation, $g_{i,end}$ and $g_{j,end}$ – their mass in the end of the operation. In case of drastic domination of component j (solvent) in the original and final solutions, i.e., when $g_{i,end}/g_{j,end} \approx C_{w,i,end}$, but $g_{i,beg}/g_{j,beg} \approx C_{w,i,beg}$, where $C_{w,i,beg}$ and $C_{w,i,end}$ are concentrations of the organic component i before and after the operation, respectively, this equation acquires this format:

$$\kappa_i = \frac{C_{w,i,\text{end}}}{C_{w,i,\text{beg}}}. \tag{1.23}$$

The solvent j is usually water. However, this equation is also valid when at concentration, the nature of the solvent j also changes. For instance, in extraction, such dominating component j before the operation is water and in the end of the operation, organic solvent, i.e., extractant. Then the component concentrating factor i is described by the ratio of its initial concentration in water and its final concentration in extractant.

Another parameter of concentrating determines the fraction of the extracted component i of its content in natural water and is called ***extraction efficiency*** or ***extraction component***. This feature is determined from the following equation:

$$R_i = \frac{g_{i,\text{end}}}{g_{j,\text{beg}}} \tag{1.24}$$

It follows from here that

$$R_i = \kappa_i \frac{g_{j,\text{end}}}{g_{j,\text{beg}}} = \kappa_i \frac{V_{j,\text{end}}}{V_{j,\text{beg}}}, \tag{1.25}$$

where $g_{j,beg}$ and $g_{j,end}$ are weights and $V_{j,beg}$ and $V_{j,end}$ are volumes of solvent before and after the operation, respectively.

The major methods of concentrating ground water organic analytes are distilling and rectification, extraction (gas, liquid and solid-phase), freezing-out, ultrafiltration and centrifuging.

Distillation and ***rectification*** or distilling with water steam is applied for extraction and concentrating volatile organic components. At that, the extent of analyte's extraction depends on volume of moisture being evaporated. This technique is utilized both for nonpolar (hydrocarbons) and amphiphilic (acids, amines, phenols and alcohols) analytes.

The division at distillation depends on the nature of the volatile component. In the simplest case with practically water-insoluble nonpolar compounds the steam pressure decreases with the increase of their molecular mass and does not change due to the presence of water. At distillation of such compounds with water the ratio of their mass in the condensate depends only on their molecular mass and saturated steam pressures at given temperature:

$$\frac{g_{i,\text{end}}}{g_{H_2O,\text{end}}} = \frac{P_i^0}{P_{H_2O}^0}\frac{M_i}{M_{H_2O}}, \qquad (1.26)$$

where P_i^0 and $P_{H_2O}^0$ are pressures of saturated steam at the given temperature. The value of saturated steam pressures ratio $a_i^0 = P_i^0 / P_{H_2O}^0$ for this reason has great significance and is called ***selectivity factor***. In case of the water steam the final concentration of the component i is determined so:

$$C_{B,i,\text{end}} = \frac{g_{i,\text{end}}}{g_{H_2O,\text{end}}} = \frac{P_i^0}{P_{H_2O}^0}\frac{M_i}{18} = a_i^0\frac{M_i}{18}. \qquad (1.27)$$

And from here the concentrating factor at distillation with water steam is equal, according to equation I-23:

$$K_i = \frac{P_i^0}{P_{H_2O}^0}\frac{M_i}{18C_{w,i,\text{beg}}} = a_i^0\frac{M_i}{18C_{w,i,\text{beg}}} \qquad (1.28)$$

These equations show that the distillation effectiveness grows with the growth in the analyte of values M_i and a_i^0. The value of the inverse final concentration,

$$\frac{1}{C_{w,i,\text{end}}} = \frac{g_{H_2O}}{g_i} = \frac{18P_{H_2O}^0}{M_iP_i^0} = \frac{18}{a_i^0M_i}, \qquad (1.29)$$

is the ***discharge factor***, which describes the weight of water steam necessary for distillation of one weight unit of the organic analyte. If the discharge factor is insufficiently high, the rectification, i.e., numerous repeating of the condensation process is used. In this case

$$C_{w,i,\text{end}} \approx K_i^nC_{w,i,\text{beg}}, \qquad (1.30)$$

where n is the number of rectifications. If the studied organic matter is decomposing when heated, the distillation is conducted in vacuum.

Soluble amphiphilic components form hydrates with H_2O. For this reason pressure of their steam in the presence of water noticeably decreases and the volatility increases with growth of the molecular mass. Due to this

$$\lg \frac{C^{*}_{w,i,beg}}{C^{*}_{w,i,end}} = \alpha_i \log \frac{C^{*}_{w,H_2O,beg}}{C^{*}_{w,H_2O,end}}, \quad (1.31)$$

where $C^{*}_{w,i,beg}$ and $C^{*}_{w,i,end}$ are weight concentrations of organic analyte in the distillate in the beginning and at the end of distillation, $C^{*}_{w,H_2O,beg}$ and $C^{*}_{w,H_2O,end}$ are weight concentrations of moisture in distillate in the beginning and at the end of distillationи, α_i is relative volatility of organic analyte and moisture in water steam. ***Relative volatility*** in this case is equal to the steam pressure ratio of the pure volatile component *i* to the steam pressure of the water steam.

This concentrating technique is used mostly for relatively well water-soluble amphiphilic components, acetone, alcohols, phenols and organic acids, including humic, etc. Concentrating of such organic components may require additional volumes of distillate. Then either steam generators are used or distilled water is added in the distillation flask.

Extraction assumes the extraction of an organic analyte from water with greater than effective absorbent, which is called ***extractant*** (*leach*). Depending on the nature of extractant the gas, liquid and solid-phase extractions are distinguished.

Dynamic gas extraction is used for extraction of most volatile organic compounds. This technique is also called gas blowdown, stripping or *purge and trap*. A gas, for instance pure N_2, is run through a measured volume of sample. The volatile organics are transferred into the gas and with it gets in the trap. This trap is a sorbent-filled metallic tube. Activated charcoal, silica jell (*Carbopack, Tenax*) and carbon molecular sieves (*Carbosieve, Carboxen*) are most commonly used as sorbents. Before analysis the organics are desorbed by heating and blowing through the trap and are transferred in the gas chromatograph column.

Liquid-liquid extraction is the extraction of organic matter with liquid organic solvents and is used mostly for semivolatile and non-volatile organic components. This is the most common division technique. Pentane (C_5H_{12}), hexane (C_6H_{14}), cyclohexane (C_6H_{12}), benzene (C_6H_6), chloroform ($CHCl_3$), dichlormethane (CH_2Cl_2), ethylacetate ($CH_3COOC_2H_5$), octanole ($CH_3(CH_2)_6CH_2OH$), petroleum-ether, etc. are used as extractants. Selection of the organic solvent is determined by its properties, purity, subsequent removal technique, stability at safekeeping and toxicity.

Usually for the extraction of semivolatile organic compounds water is alkalified to a pH higher than 11 and then the alkali-neutral fraction is extracted by dichlormethane. After this the sample is acidified to a pH below 2 and the acidic fraction is extracted by dichlormethane. Each fraction is dried up by the removal of the moisture, concentrated by distillation to 1 ml and sent for analysis.

The concentrating factor at extraction depends on values of ***distribution factor*** $\tilde{K}_i$, i.e., ratio of the concentrations of the same component in extractant and in water. In the extraction the organic analyte i is distributed between water and extractant so that the following equation is valid:

$$C_{\mathrm{w},i,\mathrm{beg}}V = \tilde{C}_{\mathrm{w},i}\tilde{V} + V\frac{\tilde{C}_{\mathrm{w},i}}{\tilde{K}_i}, \tag{1.32}$$

where $C_{\mathrm{w},i,\mathrm{beg}}$ and $\tilde{C}_{\mathrm{w},i}$ are the concentrations of analyte i in the water sample and in extractant, respectively, V and $\tilde{V}$ are interacting volumes of the water sample and extractant, $\tilde{K}_i$ is component distribution factor i between extractant and water, i.e., the ratio of its concentrations in these media under the equilibrium conditions. It follows from this equation that the analyte concentration i in the extractant is equal to:

$$\tilde{C}_{\mathrm{w},i} = C_{\mathrm{w},i,\mathrm{beg}}\frac{\tilde{K}_i V}{\tilde{K}_i\tilde{V} + V}. \tag{1.33}$$

For this reason the concentrating (division) factor at extraction is:

$$\kappa_i = \frac{\tilde{C}_{\mathrm{w},i}}{C_{\mathrm{w},i,\mathrm{beg}}} = \frac{\tilde{K}_i}{1 + \tilde{K}_i\frac{\tilde{V}}{V}}, \tag{1.34}$$

And the fraction of extracted analyte i of its initial content in the sample, i.e., the extent of extraction, is equal to:

$$R_i = \frac{\tilde{K}_i}{\tilde{K}_i + \frac{V}{\tilde{V}}} = \kappa_i\frac{\tilde{V}}{V}. \tag{1.35}$$

The higher the distribution factor value, the greater than effective the extraction. The analyte residue in water after the first extraction $C_{\mathrm{w},i,\kappa}$ is determined form the following equation:

$$C_{\mathrm{w},i,\mathrm{end}} = C_{\mathrm{w},i,\mathrm{beg}}\frac{1}{\tilde{K}_i\frac{\tilde{V}}{V} + 1} = C_{\mathrm{w},i,\mathrm{beg}}\frac{\kappa_i}{\tilde{K}_i} \tag{1.36}$$

Then after the nth sequential extraction the analyte residue in the sample will be equal to:

$$C_{\text{w},i,\text{end}} = C_{\text{w},i,\text{beg}} \left(\frac{1}{\tilde{K}_i \frac{\tilde{V}}{V} + 1} \right)^n = C_{\text{w},i,\text{beg}} \left(\frac{K_i}{\tilde{K}_i} \right)^n, \tag{1.37}$$

provided the ratio $V/\tilde{V}$ in all sequential extractions is equal.

The extraction of water-soluble organic compounds with low $\tilde{K}_i$ value (carboxylic acid, amino acid, etc.) is combined with vaporization or adsorption.

Solid-phase extraction or molecular sorption utilizes concentrating organic matter on the surface of solid matter - adsorbents. For this purpose either the adsorbent is placed within the sample and held there until adsorption equilibrium or the sample is run through a cylinder with adsorbent. Most common adsorbents are activated charcoal and macro-porous resins. Adsorption of polar compounds depends on pH of the water. They adsorbed better in an acidic medium. Desorption is conducted before the analysis by extraction in organic solvents (most often in chloroform) or in water solutions with different pH values. Drawbacks of the solid-phase extraction are its possible irreversibility as a process and oxidizing of organic matter in the presence of oxygen.

Freezing-out is a technique of mild concentrating. It is based on gradual freezing-out of moisture and concentrating of the organic components in the residual liquid solution. The process is stopped before the studied components begin transiting to solid phase. Effectiveness of the technique depends on the speed of freezing, mineral and organic composition of the water. The concentrating factor of this technique reaches 10-20.

Ultrafiltration and centrifuging are used for separating organic macromolecules with the size in excess of 0.1 μm (humic acids, polysaccharides, proteins, etc.) capable of forming colloid solutions. In this case semipermeable membranes with pores between a few and hundreds of nanometers or centrifuges with the revolution rate of up to $6{\cdot}10^4$ rpm are used.

Requirements to the conservation, concentrating or dividing of organic matter in the composition of ground water recommended in GOST P 51592-2000 are listed in Table 1.10.

In integrated studies of the composition and content of organic matter the technique of dividing and concentrating may have a very complex combination nature.

1.4.2 Analysis of Organic Substance

Methods of organic substance analysis depend on its properties, composition and nature of the assigned tasks. Often it is sufficient to estimate the general content of dissolved organic matter, less often it is necessary to measure total content of groups of its compounds, and even less often it is required to determine the nature and content of specific organic compounds. In Russia standard analyses of organic matter are regulated by GOST's: P 51209-98, P 51209, P 51232-98, P 51310-99, and P 51392-99, 27384-2002. In the USA standard analysis methods of organic compounds dissolved in water are provided in U.S. EPA Series 500 and 600. Methods of the series 500 are developed for analysis of potable water and methods of the series 600, for municipal and industrial waste water.

All analyses of organic matter in water may be subdivided in three groups: 1. determination of total dissolved organic substance content, 2. determination of total content of individual groups of organic components with similar properties, 3. determination of the content of individual organic compounds.

1.4.2.1 General Content of Organic Matter

Organic matter is represented in ground waters by micro-concentrations of a great number of compounds with broad range of volatility and capability to withstand thermal effects. For this reason it is very difficult to determine the content of organic matter as a sum of concentrations of its individual components. For a rough estimate of general organic matter content, usually the value of ***biochemical consumption of oxygen*** (BOD) is found, and for a greater than accurate estimate, the value of ***chemical oxygen demand*** (COD) or ***oxidability***, but sometimes even directly the ***content of individual biogenic elements*** (C, N, P) in the water can be found (Table 1.11).

Biochemical consumption of oxygen (BOD) describes its use by aerobic microorganisms in milligrams per 1 liter of water. The BOD value is the crudest estimate of general organic matter content. This technique (see below for details) is used in highly polluted mostly surface and waste waters with high organic content. Its reliability depends on the nature of microorganisms, composition of organic matter and also the amount of dissolved oxygen. Besides, microorganisms may also oxidize some inorganic compounds (NO_3^-, NO_2^-, etc.), and that results in overestimated BOD.

A greater than reliable estimate of the general amount of organic matter in the water is provided by the value of oxidability or chemical oxygen demand (COD). It gives milligrams of oxygen in the composition of strong

Table 1.11 Determination methods of TOC and organic nitrogen in the natural waters.

Components	Methods of analysis
TOC of volatiles and neutrals	Catalytic burning in water steam
TOC of nonvolatiles	Dry burning in empty tube
TOC total	Wet burning (GOST P 52991-2008) Instrumental methods (TOPAZ-C, TOPAZ-NC, Formacs TOC/TN, etc.)
Organic nitrogen (mostly in ammonia)	Modification of Johan Kjeldahl (1849-1900) technique.
Organic nitrogen total	Reductive destruction (modification of Krogh–Keys technique) Instrumental methods (TOPAZ-N, Primacs^{SNC}, Formacs TOC/TN, etc.)

oxidant ($KMnO_4$, KIO_3, $K_2Cr_2O_7$) utilized for oxidizing 1 liter of a solution. Methods to determine this value are provided below. In this case oxygen is also expended both for organic and inorganic deoxidants (Fe^{2+}, H_2S, NO_2^-, etc.). For this reason the general content of organic matter is estimated from a difference between COD and calculated oxidability of solution's inorganic components. Oxidability of inorganic deoxidants is determined based on water 's mineral composition from the consideration that oxygen is expended in the amount of 0.47 mg per 1 mg H_2S, 0.35 mg per 1 mg NO_2^- and 0.14 mg per 1 mg Fe^{2+}. Chloride is also partially oxidized. For this reason waters with Cl^- content of over 300 mg/l supply overestimated amounts of organic matter. On the contrary, the organic compounds, which include, in particular, nitrogen in their composition, e.g. proteins, poorly oxidize and thereby understate the organic matter content. It is necessary to take into account that in ground waters are dissolved organic components that are greater than stable to oxidizing, so this underestimation may be significant. For the determination of ground water COD it is recommended to use the strongest oxidants (KIO_3, $K_2Cr_2O_7$). It is assumed that 1 mg oxygen or 4 mg potassium permanganate ($KMnO_4$) is expended for 21 mg of organic matter.

An important parameter of total organic matter concentration is the content of organic carbon (TOC). Rarely, the content of organic nitrogen N_{org} or phosphorus P_{org} is determined. ***Organic carbon*** (*total organic carbon - TOC*) is the weight content of carbon in the composition of organic matter per unit volume of water.

Analysis requires 5 to 50 ml of water. As organic matter is unstable under the aerobic conditions and capable of decomposing to CO_2, the determination of organic carbon should be done as soon as possible after the testing. If the water sample is collected into a glass vessel, acidified by the sulfuric or orthophosphoric acid to a pH below 2 and cooled to 2-5°C, it is possible to hold it no greater than than 7 days. If the sample is frozen to minus 20°C in a polymer vessel, it is possible to hold it up to 1 month.

The TOC determination technique is based on oxidizing the dissolved carbon to CO_2 and analyzing the amount of this compound. TOC of volatile and semivolatile organic compounds is determined by oxidizing the organics in the steam at evaporation of the sample (wet burning). It takes into account the presence of inorganic carbon, which is represented mostly by carbonate ions. These ions are either preliminarily removed by acidification of the samples and then blowing through by the purified air, or preliminarily analyzed and subtracted from the total carbon content.

TOC of nonvolatile organic matter is derived at oxidizing dry residue after evaporation (dry burning). Sometimes organic matter is preliminarily extracted by chloroform at various pH values then the extractant is distilled off and the dry residue of the organic matter is burnt. Then the total content of TOC is determined as the sum of TOC for individual fractions. Methods of TOC and organic nitrogen determination in natural waters are listed in Table 1.11. Sensitivity of these techniques varies between tenths and hundredths of $mg{\cdot}l^{-1}$. Content of organic carbon is usually measured in tens of $mg{\cdot}l^{-1}$ and in deep waters sometimes reaches several hundred $mg{\cdot}l^{-1}$. In brines of oil fields TOC content may reach several $g{\cdot}l^{-1}$.

It is conventional to assume that the content of organic carbon on average is 2.67 times (ratio of carbon and oxygen equivalent masses) lower than COD values ($mg\ O_2{\cdot}l^{-1}$) and 2 times lower than total content of dissolved organic matter (M.E. Altovsky believes that contents of the dissolved organic matter is 3 or even 4 times of TOC due to the loss of its volatile portion). It is usually assumed that the contents of the dissolved organic matter is equal approximately to 2·TOC or 0.75·COD.

1.4.2.2 Content of Organic Component Groups

Analysis of ***organic component groups*** determines total content of homotypic organic compounds, for instance, organic acids, petroleum products, volatile phenols, synthetic surface-active substances, etc. Their study requires diverse methods of conservation, concentrating and chemical analysis. Most commonly used are gravimetry, titrimetry, chromatography, spectrometry and photometry. Determination methods of total concentrations are usually insufficiently sensitive. Their concentrations below tenths or hundredths fractions of a milligram per 1 liter usually cannot be determined (Table 1.12).

Organic acids are of interest for the solution of many hydrogeochemistry and ecology problems. Among them are distinguished low- and high-molecular mono-carboxylic acids, di-carboxylic acids, amino acids, naphthene and humic acids, which are substantially different in their physical and chemical properties.

General content of organic acids is estimated quite approximately by methods of water and non-water potentiometric titration. For this the solution is preliminarily acidified to pH 3.2-3.3, CO_2 is removed by boiling and the excess of mineral acids is tittered out by alkali to pH 4. Volatile organic acids are concentrated directly before the analysis by distillation with steam.

Table 1.12 Analysis methods of aggregate organic matter in ground waters.

Analyzed components	**Analysis technique**
Groups of components	
Petroleum products (GOST P 51232-98, GOST P 51797-2001, RD52.24.476–95)	IR-spectrophotometry, gravimetry, gas chromatography
Phenols (RD 52.24.488–95)	Photometry, colorimetry, gas chromatography
Phenol index (ИСО 6439–90)	Spectrophotometry
Organic acids	Titrometry, gravimetry, Spectrophotometry
Humic acids	Titrometry
Fatty acids	Titrometry, gas chromatography
Naphthene acids	Turbidimetry
Volatile aromatic hydrocarbons C_4–C_7 (RD 52.24.473–95)	Gas chromatography
Volatile chloro-substituted hydrocarbons (RD 52.24.482–95)	Gas chromatography
Chlolrorganic pesticides (GOST P 51209–98)	Gas-liquid chromatography
Volatile halogenorganic (GOST P 51392–99)	Gas-liquid chromatography
Surface-active matter (surfactant) anionoactive (GOST P 51211–98, GOST P 51232-98)	Fluorimetry, spectrophotometry

The content of organic acid is determined in moles of alkali used for raising pH from 4 to 7. The obstacles are ions of phosphoric acid, iron and aluminum. The technique's selectivity may be raised if the acids are extracted into organic solvent and tittered in this medium.

Humus acids are of a special interest in hydrogeochemistry due to their active influence on the quality of ground water and migration mobility of many metals. For their concentrating, vaporization, freezing, adsorption on cellulose anion exchangers and extraction are used. Polar organic solvents (butyl and amyl alcohols) or chloroform are used for extraction. Separation of the humic from fulvic acids is done by acidification of humic acid concentrate. The sediment of humic acids is dissolved in 0.1 n solution of NaOH. The acidified solution with fulvic acids is also alkalinized to a pH around 13. The content of either one is determined from oxidability (oximetric technique) and also by gravimetry or spectrophotometry.

Phenols hold a special place among acids dissolved in water. Most dangerous among them are volatile phenols (phenol, cresols and xylenes). For the determination of the total of volatile phenols they are first concentrated by distillation with steam and then the content is analyzed photometrically. The mass concentration in water of phenols reacting with 4-amin-antipyrine and forming with it colored compounds is called the ***phenol index***. This group includes such volatile phenols, separable through stripping by the water steam, as oxy-, methoxy-, carboxy- and sulfa-derivatives. It does not include *n*-cresol and other phenols with substituted groups. The phenol index in potable waters must be below 0.25 $mg{\cdot}l^{-1}$.

Petroleum products are taken as sum of hydrocarbons without any polar organic compounds. They are a long series of nonpolar and low polar compounds soluble in hexane, which in their totality form fractions with different boiling temperatures. Lighter fractions (gasoline and kerosene) are most volatile and better soluble in water; heavier high-boiling fractions (boiler oil, et.) are less volatile and soluble. Saturated and aromatic hydrocarbons in petroleum products are often determined separately. For this reason methods of testing, concentrating and analysis of petroleum products strongly depend on the nature of the assigned problems.

Glass vessels are used exclusively for the collection of samples. At low concentrations (0.05-0.5 $mg{\cdot}l^{-1}$) sample volume must be at least 2 liters, at medium concentrations (0.5-2.0 $mg{\cdot}l^{-1}$) – 1 liter and at high concentrations (greater than 2 $mg{\cdot}l^{-1}$) 0.5 liters is sufficient.

Gravimetry, gas-chromatography, spectral photometry and fluorescent spectrometry methods are used for the determination of petroleum products. Weight technique is possible to use for the content determination of high-boiling fractions (higher than 200° C) if their content exceeds 5 mg·l^{-1}. The IR-spectral photometry technique is applied for much broader circle of petroleum products. At the sample volume of 1 liter its sensitivity reaches 0.05-0.1 mg·l^{-1}. The sensitivity of the UV-spectral photometry usually does not exceed 0.05 mg·l^{-1}. Fluorescent spectroscopy allows recording petroleum products at the level of 0.05-0.1 μg·l^{-1}. However, this technique is sensitive mostly to aromatic hydrocarbons. Automatic analyzers of petroleum products in water (for instance, FLUORAT AE-2) are developed and used, which with fluorometry enable the determination of the contents of petroleum products in the range of 0.01 – 10 mg·l^{-1}.

GOST P 51797-2001 "Water potable. Technique of determining the content of petroleum products" recommends to determine emulsified and dissolved petroleum products in household-potable waters by the IR-spectrometry technique. For this purpose the sample is first acidified to pH below 2 and petroleum products are extracted. From the extract are removed polar compounds by running it through a chromatographic column and the obtained eluate is analyzed. The content of petroleum products in drinking water must be below 0.1 mg·l^{-1} and in water reservoirs below 0.3–0.5 mg·l^{-1}.

Surface active substances include a broad range of organic compounds, among them the most important are anion-genic. For analysis is required unfiltered sample of at least 50 ml. The duration of its safekeeping is usually no longer than 2 days. The content of surfactants is determined mostly by extraction-photometric technique. Analysis methods of these substances in potable waters are regulated by GOST 51211-98 "Water potable. Methods of determining the content of surface-active compounds". Under these regulations the content of surfactants in potable waters is determined by 3 methods intended for the determination of mass concentrations:

- Anion surfactants in the rangee of 0.015 to 0.25 mg·l^{-1} – spectrophotometry;
- Anion surfactants in rangee of 0.025 to 2.0 mg·l^{-1} – fluorometry;
- Cation surfactants in rangee of 0.01 to 2.0 mg·l^{-1} – spectrophotometry.

These techniques are based on a reaction with dyes which forms colored complex compounds soluble in chloroform. These compounds are extracted by chloroform and analyzed in extractant by photometric technique. According to SanPiN 2.1.4.1074-01, the content of anion-genic surfactants in potable waters must not exceed 0.5 mg·l^{-1}.

1.4.2.3 *Content of Individual Organic Components*

Analyses of individual organic substances, especially in waste waters, are costly and associated with the following technical difficulties.

First, it is difficult to correctly identify the studied component in a broad range of similar ones present in the water at very low concentrations. To perform this, a set of chemically pure organic components, used for the identification of dissolved components, is needed.

Second, almost completely missing is the information about the behavior of many organic components in geologic media, especially in complex conditions of anthropogenic contamination.

Third, for concentrating and purification of the analyzed components chemicals and extractants of exceptionally high purity are required.

Numerous techniques exist for the determination of the type and content of water-dissolved organic components. They enable obtaining quite reliable data about content, for instance, of carboxylic acids, phenols, different hydrocarbons, amino acids, etc. Most commonly utilized among them are chromatography (gas-liquid, thin-bedded, etc.), mass-spectrometry, photometry and luminescence. In particular for the determination of pesticides, which by their chemical properties are classified into three main groups (organo-chlorine, phosphor-organic pesticides and sym-triazine herbicides), are used methods of gas-liquid and thin-bedded chromatography and for amines – photometric methods.

Often used is a combination of different techniques, in which, after chromatographic dividing, follows the identification with mass-spectrometry, IR-spectroscopy, etc. Most commonly used is the combination of gas-chromatography with mass-spectrometry. Sometimes entire complexes are created merging chromatographs, mass-spectrometers and computers. Some such techniques are listed in Tables 1.13 and 1.14.

1.4.3 Conversion of Analysis Results

If the content of an organic analyte was determined directly in the water, it usually is expressed in weight or volume units ($mg \cdot l^{-1}$, $ml \cdot l^{-1}$), which with known molecular weight may be converted in molarities or molar fractions. If their contents or concentrations were determined in the extract or concentrate, a preliminary conversion is needed taking in consideration the concentrating factor (κ_i) or segregation extent (R_i). This conversion is performed using the following equations:

$$C_{w,i,beg} = \frac{C_{w,i,end}}{\kappa_i}, \tag{1.38}$$

Table 1.13 Analysis methods of 600's series (USA) and their purpose.

Technique number	Analysis methods	Target components
601	gas chromatography	Purified halogen-containing
602	gas chromatography	Purified aromatic
603	gas chromatography	Acryl aldehyde and acrylonitrite
604	gas chromatography with flame ionization detector	Phenols
605	High-efficiency liquid chromatography	Benzidins
606	gas chromatography	Phtalate ether
607	gas chromatography	Nitrozamin
608	gas chromatography	Organochlorine pesticides and Polychlorinated biphenyl
609	gas chromatography	Nitroaromatic and iso-phoric isophorone
610	gas and high-efficiency liquid chromatography	Polycyclic aromatic hydrocarbons
611	gas chromatography	Haloethers
612	gas chromatography	Chlorinatedhydrocarbons
613	gas chromatography / mass-spectrometry	2,3,7,8-TCDD (dioxin)
624	gas chromatography / mass-spectrometry	Purified organics
625	gas chromatography / mass-spectrometry	Organics extracted by acids and bases (neutral)

or

$$C_{w,i,beg} = \frac{g_{i,end}}{R_i V} = \frac{\tilde{V}}{V}\frac{\tilde{C}_{w,i}}{R_i}. \quad (1.39)$$

Here $g_{j,end}$ is the final weight of component i after concentrating or dividing in the analyzed sample. As the second equation shows, in order to be able to use the segregation extent factor it is necessary to know V – water volume for extraction, $\tilde{V}$ – volume of the extractant or concentrate

Table 1.14 Organic matter determination methods for drinking water quality control under GOST P 51232-98.

Organic components	**Acnalysis methods**
Benzene	Colorimetry, photometry, gas-liquid chromatography
Organochlorine pesticides, GOST P 51209–98	Gas chromatography
Volatile halogenorganic compounds, GOST P 51392–99	
Lindan (γ-isomer HCH), (GOST P 51232–98, GOST P 51209)	Gas-liquid chromatography
DDT (sum of isomers) (GOST P 51232–98)	
2,4-Д (2,4-dichlor-phenoxyacetic acid) (RD 52.24.438–95)	
Tetrachlorcarbon (GOST P 51232–98)	
Benz(a)pyrene (GOST P 51310–99)	HPLC with fluorimetry
Product chemical detoxification sovtol, Methodical instructions 4.1.1091–02	High-efficiency liquid chromatography

* Valid until the approval of the corresponding State standard.

obtained at preparation. After that weight concentrations of individual organic components may be converted into molarities or molar fractions using the equations listed for mineral or gas components.

Partial pressure is the exception. Most organic compounds in pure form under normal conditions are in the form of liquid and steam. For this reason their partial pressure at low values should be determined, in compliance with Raoult's law, from this equation:

$$p_i = P_i^0 \cdot C_i, \tag{1.40}$$

where C_i is molar fraction of the component i in water, P_i^0 is pressure of the saturated steam of the same component at given temperature.

1.5 Substances in the Dispersed State

In ground water, as in any natural water, not only dissolved but also suspended matter is present. Each particle of such matter has certain separation surface, which restricts its size, and differs from the water solution in its composition. The matter, which may be removed from the solution by filtration and/or centrifuging, is considered ***suspended solids*** and is in the ***dispersed state.***

This matter hinders the light penetration and makes the water turbid, nontransparent. It possesses huge specific separation surface with water solution and for this reason actively participates in mass-exchange with water (adsorption, solution, precipitation, metabolism, etc.) Behavior and role of this matter in composition of the ground water depends on the nature, size, composition and amount of its particles.

The nature of these suspended particles may be solid minerals or organics (detritus), or liquid (drops of crude oil, petroleum products, lubricants, etc.) or even gas as bubbles of 1 to 30 μm and greater. Even some individual macromolecules of the dissolved organic matter (for instance, humic) may be so great as not to be able to get through the filter (Figure 1.10). Besides, various live microorganisms (protozoa, bacteria, viruses, etc.) may float in the water. As a whole the natural water is simultaneously the suspension (contains solid matter) and emulsion (contains bubbles and drops).

All suspended particles in their nature are subdivided, first of all, into passive inert and active living organisms.

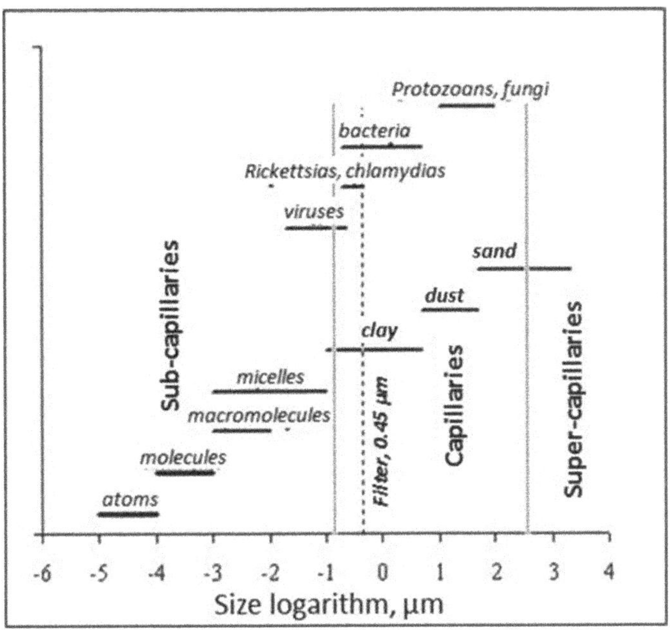

Figure 1.10 Comparing the size of dispersed matter with the size of dissolved components and pores.

1.5.1 Inert Suspended Particles

The size of inert suspended particles depends on velocity of motion and density of waters. In caves flow velocity may be quite high in order to transport sand or gravel in suspended state. At usual velocity of filtration in porous media the underground water is capable of washing off only tiny clay particles from loose rocks causing mechanical suffusion.

If a water sample is left in peace, the heavy inert matter will precipitate on the bottom but the light one will surface or even volatilize. Gas bubbles float, sometimes form foam and are lost in the atmosphere. Most undissolved organic matter (oil or petroleum product fractions, heavy hydrocarbons, lubricants, etc.) will accumulate on the surface. Its presence may cause disagreeable smell and after-taste. It makes water unsuitable for use in food. Water with elevated content of organic matter is not recommended even for making concrete.

Buoyant emulsion should be separated by carefully pouring in a clean vessel. If needed, nonvolatile organic matter may be separated by wet extraction with diethyl ether. Diethyl ether vaporizes, the residue is weighted and sent for the chemical analysis. Volatile organic compounds at such testing may be lost.

Particles of solid matter mostly precipitate on the bottom. Velocity of their immersion is proportionate to the square of their diameter. Part of the precipitate, which forms in the first 5-6 hours of settling-down and has the size of greater than than 10^{-2} mm, is called ***coarse suspension***. Elutriated water is run through a filter with pores 0.45 μm. Dispersed matter accumulated on the filter has the size 10^{-4}–10^{-2} mm and is called ***fine suspension***. It is mostly particles of clay and organic matter (detritus). Coarse and fine suspension is described by total weight content and grain size, and sometimes also mineral composition.

Particles, which got through the filter with hole size 0.45 μm are attributed to colloids. Its individual particles with the size 10^{-4}–10^{-6} mm (10^{-3} to 10^{-1} μm) are called ***micelles***. These particles include about 10^3–10^9 atoms, they are subjected to Brownian motion, poorly precipitate and are distinguished with difficulty from dissolved macromolecules. They are represented by silica ($SiO_2 \cdot H_2O$), iron, aluminum and other metal hydroxides and oxides. They may include in their composition detritus and living organisms. Micelles possess the highest specific surface, actively interact with dissolved components, and together with them form solutions, which are called ***colloidal solution***, ***sol*** or ***hydrosol***. Colloid differ from genuine solutions mostly in the number of micelles. Particles in genuine solutions have size no greater than 10^{-6} mm and cannot be discovered by optical methods. In colloid solutions micelles are large enough to disperse light ("Tyndall effect"). At very high concentrations the micelles interact between themselves forming brittle 3D structures with some properties of solid matter, which is called ***gel***.

Suspended inert matter in potable waters is unacceptable. Under the requirements to the composition and properties of household-drinking and cultural and general purposes waters, the content of suspended particles must not exceed 0.25 $mg{\cdot}l^{-1}$ and 0.75 $mg{\cdot}l^{-1}$ respectively. They make water abrasive for tubes, pumps and turbine blades. In the water used in some technological processes their content cannot be higher than 1 $mg{\cdot}l^{-1}$, and 50 $mg{\cdot}l^{-1}$ or higher is rarely accepted.

1.5.1.1 Methods of Study

Suspension may be foreign, formed due to testing. In particular, it may be captured at pumping out a well, especially when drilling solution is used, or when filtration rate at bottomhole is too high. For this reason samples of ground water from wells are better collected after a lengthy pumping out at low rate in order not to capture the bottomhole suspension. Besides, temperature and pressure decline, degassing and interaction of water with O_2 may result in the formation of new suspended particles (calcite, hydroxides of Fe, Mn and other heavy metals). A foreign suspension in a sample may noticeably affect the cation content, first of all in heavy metals, in the process of safekeeping and sample preparation for laboratory analysis. In order to eliminate this, the water before the sample collection, as a regular rule, is filtered through a membrane filter with pore size 0.45 μm.

The amount of suspension is determined by the weight technique in grams per liter of the solution. The content of coarse suspension is determined directly at the place of testing after having allowed the water to settle. Thin suspension is separated by water filtration through a paper or membrane filter or by centrifuging. Most complete separation of suspended matter is achieved with the use of membrane filters. The filter should not introduce additional contamination and hamper the solution of assigned problems. For this reason before the application it is thoroughly washed through and if needed is sterilized. Filtering is not applied only to the samples intended for gas and easily volatile component analysis, but also to well adsorbed dissolved micro-components. In individual cases samples with suspension are conserved.

Gravimetric composition of coarse and thin suspension may be determined by usual methods used for loose ground. For studying the grain size composition of micelles, ultra-filtration systems are sometimes used, in which the sample is run through a number of decreasing size hole millimicron filters. Knowing the volume of water run through this system and mass of colloids captured by each filter, it is possible to determine concentrations of different size micelles. Mineralogy of particles captured on filters is determined with help of scanning electron microscopes. For these purposes sometimes the technique of light-dispersing laser is used, which also provides information about grain size composition and colloid concentration.

1.5.2 Living Organisms

The living aggregation of ground waters is its ***biocenose***. The largest living organisms (fish, small crustaceans and even humans) may be encountered only in ground waters of caves. The bulk of living **organisms** in ground waters are microorganisms, namely, some protozoans, algae, microscopic fungi, bacteria and noncellular viruses. All these organisms are very small, visible only under microscope and capable of penetrating through rock fractures and pores (Figure 1.10). The geologic medium and ground waters serve for most of these microorganisms as the ***biotope***, i.e., their natural habitat.

The *protozoans* and *fungi* are the largest organisms (amoebas have the size of 15 μm to 1 mm), which feed mostly on ready organic matter. Their mobility is restricted by permeability of water horizons. *Viruses*, smallest (0.015-0.35 μm) noncellular particles, are comprised of the aggregation of genes (DNA or RNA) within the protein shell and can propagate only in living cells. They need a living organism. For this reason in geologic medium mostly bacteriophages - viruses living in bacteria are preserved longer. The main object of biochemical studies in ground waters are *bacteria*, small single-cellular organisms capable of ingesting both organic and mineral matter.

Most bacteria are 0.5 to 5 μm in size but some species may reach the thickness of 10 μm and length of 20 μm. This means that bacteria are capable of penetrating through pores and micro-fractures in sediments and rocks.

In the composition of the reviewed microorganisms 75-85% of the weight is moisture but their elemental composition is little different from the elemental composition of organic matter. This composition continuously changes. The entire process of ingesting nutrients, their processing and removal of waste products is called ***metabolism***. This metabolism determines the mass-exchange between living organisms and underground water.

Two independent disciplines occurred in studies of microorganisms in ground waters: the **sanitary** dealing with pathogen microorganisms, and **biochemical**, which studies the role of microorganisms in formation of ground water composition. These disciplines have completely different objects and methods of study.

1.5.2.1 Pathogen Microorganisms

The sanitary discipline oriented toward the search and study of life conditions of pathogen microorganisms, mostly bacteria and viruses, emerged, obviously, in XIX century.

In 1854 in the overpopulated central London a cholera epidemic flared up. It was believed at the time that illnesses of such type were associated with malign effluvium and fogs. In connection with that cholera maps were published where the areas of maximum number of illnesses indeed gravitated to riverside lows. However, anesthesiologist John Snow suggested that cholera is carried not by "fogs" but by drinking water. In 1854 he plotted on the map all cases of the illness and found that all sick people took the water from the well at Broad Street of the Thames terrace. Snow came to the meeting of the parish abbot and recommended taking away the handle of an ill-fated pump at this well. Now in the area there is a pub, "John Snow," where a copy of Snow's map and other information related to the epidemic are kept, and a granite stone and memorial plaque serve as reminders of the place where that ill-fated pump was located.

Currently numerous disease-causing microorganisms are identified as capable of spreading in underground water.

An amoeba (*Entamoeba histolytica*), up to 700 μm in size, may cause amoebiasis, an infusorium (*Balantidium coli*), 50-90 μm in size – balantisiasis. These Protozoans multiply by mitosis (cell division), and under unfavorable conditions they form ***cysts***, temporary forms with a sturdy shell and the size of around 14-50 μm, which provide for transition from one organism to the other through the external medium.

Bacteria carried in water may cause cholera (*Vibrio cholerae*), anthrax (***bacilli:*** *Bacillus anthracis* – a rod 5-10 μm long), botulism and tetanus (***Clostridia***: *Clostridium botulinum* and *Clostridium tetani,* respectively), wound fever or glanders (***Pseudomonadaceae***: *Pseudomonas aeruginosa, Pseudomonas mallei*), typhoid fever and paratyphoid (***Salmonellae***: *Salmonella paratyphi, Salmonella schotmuelleri, Salmonella hirschfeldii, etc.*), tularemia (*Francisella tularensis*, size 0.2-0.7 μm), dysentery (*Shigella sp.*), different forms of shakes and unpleasant forms of diarrhea (*Dyspepsia coli*). Some bacteria, in particular bacilli and clostridiums, are spread by

Table 1.15 List of pathogen and opportunistic microorganisms whose presence in water makes it potentially infectious if drunk and used for washing, bathing, etc. (compiled based on the WHO list).

Protozoans	Entamoeba histolytica
	Giardia intestinalis
	Cryptosporidium parvum
	Dracunculus medinensis
Bacteria	**Cholera vibrio**
	Salmonella
	Shigellas (originators of dysentery)
	Yersinia enterocolitica
	Pseudomonas aeruginosa
	Enteropathogenic collibacilus
	Aeromonada Spl.
	Campylobacter Eyuni and Coli
Viruses	Adenoviruses
	Enteroviruses
	Virus of hepatitis A
	Enteroviruses of hepatitis neither A nor B, hepatitis E
	Norwalk virus
	Rotavirus
	Small round viruses

spores with drastically lowered metabolism and stable within a broad range of temperatures. Getting under favorable conditions their spores germinate giving life to new bacteria.

Viruses are responsible for the spread of infectious hepatitis (jaundice), poliomielitis, foot-and-mouth-disease, shakes, intestine diseases, etc. Pathogen and opportunistic microorganisms whose presence makes the water potentially infectious are listed in Table 1.15.

However, almost all pathogen microorganisms are allochtonous toward the geologic medium and do not affect its composition. Their natural habitat, their biotype, is in a living organism. An exception is the tetanus originator *Clostridium tetani* whose biotype is soil, but humans are infected through an open wound.

In just the human colon per 1 g of content is about 30–40 billion of microorganisms. Among them may be intestinal protozoans, bacteria and viruses. Coliform bacteria dominate; they are non-spore forming rods and oval streptococci, 0.6–1.0 μm in size. A single human with only the stool releases into the environment up to 20 trillion microorganisms.

Outside of living organism microorganisms are without food and relatively rapidly die of hunger. Cysts of Lamblias are capable of surviving in the external medium up to 3 months. Some malignant bacteria can exist in ground waters for almost a year and their spores much longer. Thus, spores of bacilli and Clostridium are much greater in the resistance to unfavorable physical and chemical factors than the bacteria themselves. For this reason clostridium spores sometimes are used as an indicator of drinking water quality. The greatest danger in this respect is presented by the anthrax bacillus whose spores may be preserved for many years. For this reason the anthrax animal burial sites numbering in Russia about 35 thous. are of a great danger.

Viruses cannot exist outside of a living organism. But in underground waters ***bacteriophage*** viruses (width 0.06–0.09 μm, length 0.2 μm) may be encountered capable of infecting bacteria and reproducing within them. Malignant viruses outside of organism may be preserved up to a half year. In this time they may migrate over a distance of several hundred meters.

Pathogenic microorganisms are encountered in ground waters much rarer than on the surface. Nevertheless, even their slight presence may be of great danger for humans and animals. Their content determines ground water epidemic suitability for drinking.

1.5.2.1.1 Parameters of water sanitary qualities

As pathogenic microbes do not have any effect on ground water composition and properties, it is very difficult to discover and determine their species in geologic medium. Sanitary-micro-biologic studies are directed, first of all, on the determination of the simplest indirect parameters of their presence in locations of possible anthropogenic, mostly fecal contamination. Such general parameters are total microbial number and content of indicator microorganisms.

The ***total microbial number*** (*colony count*) is the number of colonies (visible at 2-x magnification), which are formed by microorganisms from 1 ml of water on a substrate (meat-and-peptone agar) in Petri dishes after a certain incubation period (usually 24 or 48 hours) at an assigned temperature. This number provides a rough relative estimate of the general number of microorganisms in 1 ml of water.

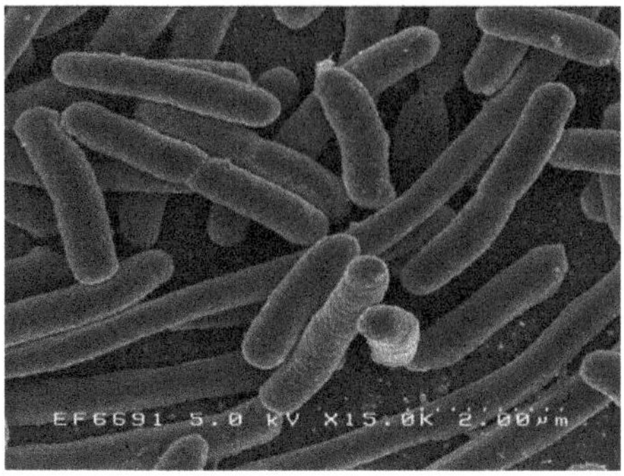

Figure 1.11 Coliform bacteria, i.e., in form of rods, ***Escherichia coli*** at great magnification. (http://dic.academic.ru/dic.nsf/ruwiki/30644)

The ***indicated microorganisms*** are groups of bacteria indicating the presence of anthropogenic, first of all, fecal contamination of waters. Usual indicators are so called coliform bacteria, not forming spores, which in terms of preservation period in geologic medium are closest to pathogenic. Their nature determines length of time; and their number, the extent, of contamination. Among them is identified a group of ***thermotolerant coliform bacteria*** with the indications of general coliforms, but capable of fermenting lactose at 44 – 45 °C. These bacteria may be rapidly discovered and for this reason play an important, although secondary role in the estimation of water pollution by fecal microorganisms. As some thermotolerant coliforms are not associated with fecal contamination, the most reliable indicator is ***collibacilus***, i.e., *Escherichia coli* (better known as *E.Coli*).

Finding *Escherichia coli* (Figure 1.11) in water indicates that this water may be polluted with fecal matter and may contain pathogenic microorganisms. Their content is estimated in units of coli-titer or coli-index. ***Coli-titer*** is volume of water in milliliters, which is needed for a single enteric bacterium. ***Coli-index*** is the number enteric bacteria in one liter of water. The water is considered healthy if the coli-titer is greater than 100 ml, quite healthy at 10-100 ml, dubiously healthy at 1–10 ml, unhealthy at 0.1–1 ml, etc. Drinking water must not have coli-titer less than 300 ml or coli-index greater than 3.

Another indicator microorganism is ***sulphate-reducing Clostridium***. Clostridium spores are capable of existing in the water much longer than coliform organisms and they have greater resistance to dehydration. Their presence in disinfected water may indicate its insufficient cleanup, therefore, that pathogenic microorganisms resistant to dehydration might not have been destroyed. The sulphate-reducing Clostridium usually is used for discovering periodic or old contamination. But not always they have fecal origin as many other sources may exist.

Indicator microorganisms of viral contamination are bacteriophages capable of infecting *Escherichia coli*. They are called ***coliphages*** and due to the similarity with human enteric viruses and strong resistance compared with indicator bacteria groups are considered as parameters of possible viral water contamination.

At last, a direct parameter of dangerous ground water contamination is the presence of pathogenic microorganisms proper. Among them the greatest attention is attracted to:

1. Lamblia cysts;
2. Vibrio's, facultative anaerobes, some of which (*Vibrio cholera* and *Vibrio parahaemolyticus*) cause cholera in humans and vibriosis in animals;
3. mostly aerobic Pseudomonaceae are very common in soil and water, some of them (*Pseudomonas aeruginosa, Pseudomonas mallei*) cause wound fever or glanders;
4. spores of anaerobic clostridium.

Hygienic standards of microbiologic and parasitological parameters considered in SanPiN 2.1.4.1074-01 (2002) are listed in Table 1.16.

Table 1.16 Sanitary standards of potable water qualities by micro-biologic and parasitological parameters (SanPiN 2.1.4.1074-01).

Parameters	Units	Standards
General microbial number	Number of colony-forming bacteria per 1ml	He greater than 50
General coliform bacteria	Number of bacteria per 100 ml	None
Thermotolerant coliform bacteria	Number bacteria per 100 ml	None
Colifages	Number of plaque-forming units (pfu) per 100 ml	None
Spores of sulphate-reducing Clostridium	Number of spores per 20 ml	None
Lamblia cysts	Number of cysts per 50 ml	None

1.5.2.2 Biochemical Microorganisms

The biochemical discipline studies a direct participation of microorganisms in the formation of composition and quality of ground waters, oils, gases and rocks. This discipline studies life-sustaining activity of the bacteria for which geologic medium serves the biotype. It contains everything necessary for their life and multiplication: the water, biogenic elements and sources of energy.

These micro-organisms are capable of existing within a broad range of temperature, pressure and water salinity. Their influence is manifested, first of all, by the acceleration of chemical reactions of organic matter oxidizing and decomposition. They are capable of decomposing practically any natural compound and thereby facilitate the circulation of biogenic elements on Earth. The study of these microorganisms includes not only determination of their varieties, their species and groups, but also study of biochemical processes, in which they take part.

1.5.2.3 Methods of Study

It is very important in micro-biologic studies to maintain sterility of the utilized equipment. It is achieved by ***sterilization, i.e.,*** complete destruction of all microorganisms regardless of their form of existence and malignancy. All vessels for testing, devices and used substances must be cleaned of microorganisms. In connection with this the vessels for samples preliminarily are carefully sterilized by high temperature (boiling for 10–15 minutes in dry-air sterilizers, in autoclaves) or by ionizing irradiation.

For samples sterile glass vessels are used, capacity 0.5 liter, with a ground stopper. Samples for pathogenic micro-flora must be 1–2 liters. The bottleneck with the stopper is covered by a sterile paper cap, which should be tied around with a thread and a twine. The water should be studied no later than 2 hours after collection of the sample. As an exception, an increase of the safekeeping period in the summer to 4 hours, and in winter to 6 hours, is accepted. If cooled to 2–10°C the samples are possible to hold up to 6 hours. For the waters with toxic metals (Be, Hg, Cd, Te) at mass concentration greater than 0.01 $mg{\cdot}dm^{-3}$, in vessels, prior to their sterilization is added 0.3 cm^3 of 15% solution of NTA (nitriltriacetic acid) per 500 cm^3 of the sample.

The mandatory element of a micro-biologic analysis is inoculating and propagation of microorganisms in a special nutrient solution. The inoculating is transfer of some volume of water samples in the nutrition medium, the propagation is multiplication of microorganisms in this medium under favorable conditions (temperature, pH). The nutrient medium is a sterile substrate from the mix of matter necessary for the growth and division of the analyzed microorganisms. This medium may be natural or artificial, liquid or solid, with addition of agar polysaccharide. The nutrient medium is selected so that only studied organisms could multiply. The direct study of microorganisms is performed after some period by the use of various optical and electron microscopes. In some cases they are colored. Reliable study of microorganisms is possible only with their pure cultures, i.e., with insulation of individual species.

Methods of sanitary-bacteriologic analysis are regulated by GOST's 24849-81 "Water potable. Field methods of sanitary-micro-biologic analysis" and 18963-73 "Water potable. Methods of sanitary-bacteriologic analysis".

For the determination of general microbial number, 1 ml of water is inoculated by a pipette onto a nutrient agar and held 24±2 hours at 37±0.5°C. The presence of bacteria is determined by the formation of colonies around each of them (Figure 1.12). Usually these colonies are possible to see with the naked eye. However, the count is conducted at the magnification 2-x to 5-x.

Several techniques exist for the determination of the number of enteric rods. The most recommended are methods of membrane filters and fermentation. In the first case bacteria are preliminarily concentrated from water on a membrane filter. With the fermentation technique the concentrating is eliminated. In both cases inoculation on a solid Endo medium and fermentation at 37±0.5°C for 18–24 hours is conducted. Colonies of enteric rods are dark red, pink and transparent. They are colored, the grown colonies are differentiated and counted. The number of colonies corresponds with the number of enteric rods. Results

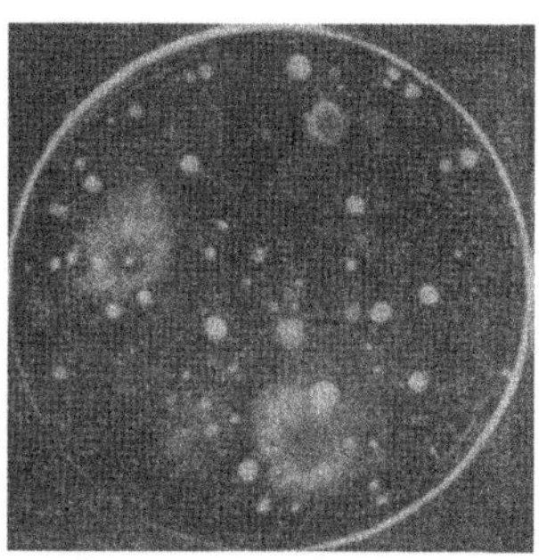

Figure 1.12 Colonies of bacteria on the medium with beef-extract agar in Petri dish.

are expressed as value of the coli-index, i.e., the number of enteric rods per 1liter of water:

$$\text{Coli - index} = 1000\frac{N}{V_w}, \tag{1.41}$$

where N is the number of discovered enteric rod colonies, $V_{w.}$ is water volume at inoculation. Coli-titer is equal to 1000/coli-indices. Sensitivity of the calculation technique is equal to one rod. For this reason, if enteric rods are not found under microscope, their number is described by value <1.

Example 1.6 At the inoculation of 10 ml of water are discovered 13 enteric rods. In this case coli-index is:

$$Coli-index = 1000\frac{13}{10} = 1300.$$

At the inoculation of 200 ml, enteric rods are not found. What is the value of the coli-index? The absence of the enteric rods means that their number in 200 ml is less than 1. Therefore,

$$Coli-index < 1000\frac{1}{200} < 5$$

At the inoculation of three volumes of water, 100 ml each, 3, 2 and 0 colonies are fermented. What is the coli-index? Analysis showed that 300 ml samples gave 5 enteric rods. For this reason:

$$Coli-index = 1000\frac{5}{300} = 17$$

Testing for the virus is conducted in living cells of bacteria. Viruses at multiplication form in the samples for testing spots of dead bacteria among the living. Each spot corresponds with one virus. For this reason virus' content is determined in 1 liter of water by the number of enteric rods dead zones (plaques) after 18±2 hours at 37±1°C on the nutrition agar.

The identification of individual species of pathogenic microbes in the waters is difficult due to their scarcity and propensity to genetic mutations. There are luminescent-serological and adsorption-serological methods, which enable their identification. But these studies are conducted only in case of epidemic danger.

1.6 Properties of Ground Water

Water properties in this case are understood as any manifestation of natural water qualities as a solution as a whole. These properties depend on the conditions of water existence and on the method of affecting it.

Underground water reacts very sensitively to the changes in the environment. At the extraction from the subsurface to the surface, and especially with placing under analysis conditions, ground water properties and also composition substantially change. This is caused, first of all, by the change in pressure and temperature but sometimes by degassing or sediment precipitation. Even if loss of the matter is excluded, only elemental composition of waters remain unchanged. For this reason it is important to distinguish ground water properties in two different states:

- subsurface, *in situ*, in the conditions of their natural existence where ground waters often are called formation waters, and
- in the conditions of their study on the surface, which are called normal, standard or simply atmospheric.

These states practically never coincide and the difference between them was so much greater when the studied underground water was positioned much deeper. In practice, hydrogeochemistry studies mostly deal with properties determined under atmospheric conditions. These properties will be reviewed here. Properties of waters as well as its composition in reservoir conditions are determined relatively rarely by calculating or experimentally.

Among atmospheric or laboratory conditions normal and standard conditions have special significance. ***Normal conditions*** (*standard temperature and pressure, STP*), according to the International union of theoretical and applied chemistry (*IUPAC*), (*Standard temperature and pressure,* STP) are considered conditions with the atmospheric pressure of 101,325 Pa (760 mm Mercury) and air temperature 273,15 K (0°C). ***Standard conditions*** *(Stndard Ambient Temperature and Pressure,* SATP) are conditions with temperature and pressure of ambient medium 25°C and 100 kPa, respectively. In actuality, water properties and composition are studied in various conditions that are close to standard, often in a laboratory.

Properties of ground water under atmospheric conditions have, as a regular rule, practical significance as they exactly determine its commercial value as economic deposit and the expediency of its use. Properties of ground water in reservoir conditions have strictly hydrogeochemical significance, as they are very important for solving issues associated with the formation of their composition or of economic deposits, and for estimation of water reserves as an economic deposit.

Besides, water properties depend on the method of affecting them and their response reaction to this action. Depending on this, organoleptic (sensory), balneological, chemical and physical properties are distinguished.

From the time immemorial, water quality was determined only with the help of sense organs, which allowed distinguishing them by color, smell and taste. This most rapid and simple technique of estimating water qualities is also used currently. Water properties determined by this technique, i.e., by sense organs are called ***organoleptic properties***. Observations and accumulated experience enabled humans to identify waters with curative qualities. Properties of waters, which are manifested in its effect on the human organism, i.e., determine its curative qualities, are called ***balneological***. Initially they were determined only empirically, as a result of multiannual observations. Legends, lore and traditions held and accumulated the information about favorable effect of some waters on human organism. This formed concepts of holy waters, which cured the people of sterility, different illnesses or even rejuvenated them. These water quality parameters, as well as organoleptic, were studied and used by humans way before the nature of water was established and its composition studied. Eventually successes in chemistry and physics resulted in methods, which allowed replacing organoleptic methods by more reliable and accurate ones. They discovered new qualities of water, which not only distinguish it from other matter but also enable distinguishing waters between themselves. These qualities are subdivided into physical and chemical properties of waters.

1.6.1 Organoleptic and Balneological Properties

Organoleptic properties describe the capacity of water to directly affect organs of sense (vision, sense of smell, taste and sense of touch). They include color, smell, taste and temperature. These parameters are unconsciously noticed by any animal and a cautious human evaluates them every time he/she intends to drink the water, as they are first estimates of its qualities. Moreover, quality of many beverages is determined by organoleptic properties of the used ground water.

For instance, celebrated kinds of English beer "Burton-on-Trent" are prepared from the hard water of local Upper Triassic sandstones. It was estimated as a joke that lovers of the Burton beer consume annually at least 159 kg of gypsum. Similarly, the quality of the best Austrian beer Gösser is determined to a significant extent by the properties of underground waters, which emerge as springs in Alps from the depth of around 200 m. Quality of Scotch whiskey depends to a large extent on the composition of used water. A view even exists that whiskey quality substantially changes if the used water first ran through granites and only then through peat, not the other way around.

Organoleptic properties include: transparence, turbidity, color, smell, taste and warmth. Initially they were determined directly by sense organs, i.e., eyes, nose, tong and hands, but their parameters were strictly subjective. In order to eliminate this subjectivity many of them got measured by instrumental methods (thermometer, turbidimeter, colorimeter, etc.) For this reason the attribution of some properties to organoleptic is associated to a great extent with tradition.

Turbidity and transparence represent the capacity of water to disperse and let through the light. These parameters are in inverse proportion relative to each other. Suspended particles increase water turbidity but decrease transparence. For this reason turbidity and transparence are used, first of all, for visual estimate of the content and nature of suspended matter.

By these parameters the water is subdivided into transparent, weakly opalescent, opalescent, slightly turbid, turbid and strongly turbid. Solutions are called ***opalescent*** when they are transparent in incident light but turbid in the reflected. Turbid solutions are non-transparent both in the incident and in reflected light.

Currently, the solution's turbidity is determined instrumentally by turbidimeter. This technique is based on photometric turbidity comparison of the studied water with turbidity of the standard suspension with different contents of uniform suspension. The determination is performed both in the incident and reflected light at different illumination angles. As the standard suspensions silica, kaolin, formazine and other thinly dispersed matter are used. The formazine suspension gives the greatest reproducibility and possibility of long safekeeping. Results of instrumental turbidity measurements are expressed in milligrams of matter of the standard suspension per 1 liter. Using kaolin the result is expressed directly in $mg{\cdot}l^{-1}$, and with the application of formazine, in $EM{\cdot}l^{-1}$ (EM is unit of turbidity) or directly in EMF ($mg{\cdot}l^{-1}$ by formazine). 1.5 $mg{\cdot}l^{-1}$ of kaolin corresponds with 2.6 $EM{\cdot}l^{-1}$ of formazine.

Turbidimetric determinations are especially important for waters with variable composition of thinly dispersed admixtures. If a sample is not filtered preliminarily, not only colloid but also more coarsely dispersed particles will account for turbidimetrically. Requirements to turbidity determination technique may be found in GOST 3351-7. Under hygienic requirements to drinking water quality the turbidity should not exceed 1.5 $mg \cdot l^{-1}$ by kaolin.

Transparence depends on the suspension and water color. Its measure is the height of the water column through which it is possible to see a white disk, 20-30 cm in diameter sunk in the water (Secci disk) or to distinguish the font of certain size and type on a white paper. The Secci disk is used for field estimates of surface water transparence. Its measurement is expressed in meters of depth.

In laboratory conditions the water is poured into a transparent measuring cylinder with a flat bottom. Under the cylinder at a distance 4 cm from the bottom is placed a font with the letters 2 mm high and 0.5 mm thick. Then the water is poured until this font can no longer be seen from the top through the water layer. The height of the water column expresses the degree of transparence. The results derived in a laboratory from the font are expressed in centimeters with indication of measurement conditions. These values are measured with the accuracy of up to 0.5 cm. At the water transparence less than 3 cm of water usage is restricted.

Chromaticity serves as a second visual feature of the natural water and determines intensity and nature of its coloration.

Two kinds are distinguished: "genuine color" caused only by the dissolved matter and "apparent color" caused by the presence in water of colloid and suspended particles whose content noticeably depends on pH values. For this reason if suspended particles are present, the sample is filtered through the paper filter. Besides, in some highly-colored waters the color may dim during safekeeping under the influence of biological processes. For this reason color determination should be done within a reasonable period of time, no later than 6 hours after collecting the sample.

Visual determination of chromaticity is conducted against white background in the layer of transparent filtered water of at least 10 cm high. Under visual evaluation the water may be colorless, greenish, yellow, fulvous, etc.

Instrumental determination of chromaticity is based on the use of photometers for comparing color samples with the color of standard solutions. Normalized technique under GOST 52769-2007 is based on visual comparison of the analyzed water color with the artificial standard color scale.

Currently the chrome-cobalt or platinum-cobalt scale is used. The chrome-cobalt one is formed by a series of potassium bichromate ($K_2Cr_2O_7$) and cobalt sulphate ($CoSO_4 \cdot 7H_2O$) solutions. The commonly used platinum-cobalt scale is represented by a series of solutions with different contents of chloroplatinic acid ($H_2PtCl_6 \cdot 6H_2O$) and cobalt chloride hexahydrate ($CoCl_2 \cdot 6H_2O$). These solutions are known as "Hazen color standards". Their use is described in GOST 29131-91 "Liquid chemical products. Color measurement technique is in Hazen units or HU (platinum-cobalt scale)." Under this scale water chromaticity varies within the range of 0 to 1,000°. 1 degree corresponds with chromaticity of the solution of 1 mg of platinum per liter. In recent times automatic colorimeters were designed, which enable rapid and reliable determination of water chromaticity by basic color scales including Hazen scale.

Water chromaticity is usually in the range of zero to several hundred degrees on Hazen scale and depends on the composition of the dissolved matter. The strongest effect on the ground water color is from the presence of humic matter and Fe^{3+} compounds. Waste water may also noticeably change chromaticity of ground water.

Color characterizes water quality. In the household, water color is undesirable not only due to esthetic considerations. The presence of coloring components damages water and food taste, and hinders the operation of devices and achieving whiteness when washing linens. Water color is undesirable in making food products, beverages, ice, textile and photography. Max acceptable value of chromaticity in the waters used for potable purposes is 35° on the platinum-cobalt scale. HHS USA believes that chromaticity of potable and culinary waters should not exceed 20°.

Odor is the capacity of the water to affect the sense of smell. There is no smell in pure water. Its presence, as a regular rule, is associated with volatile, mostly gas and organic components (H_2S, petroleum products, chlorine, phenol, etc.) in water composition. The smell may be associated with decomposition of organic matter by microorganisms.

Noticeable smell is distinguished by quality and intensity. By the quality smell may be hydrogen sulfide, swampy, fishy, putrescent, etc. Smell intensity depends on the temperature. By the intensity is distinguished very weak, weak, noticeable, clear and very strong smell. Smell intensity is measured in balls determined by GOST 3351-74 (Table 1.17).

It is possible to determine smell intensity quantitatively as the extent of dilution of the samples by the water without smell until the complete removal of smell. Minimum concentration at which the smell becomes noticeable is called ***odor threshold***. Its value is determined by sequential dilution of samples by the water without smell until the smell totally disappears. Value of the smell odor threshold is subjective as the sensitivity of olfactory organs is different in different people.

The extent of dilution at which total deodoration is reached is the ***threshold odor number***.

$$S_{o,i} = \frac{V_{test} - V_{H_2O}}{V_{test}}, \tag{1.42}$$

where $S_{o,i}$ is smell threshold odor number for volume concentrations, V_{test} is volume of analyzed water samples (with smell) in ml and V_{H_2O} is volume

Table 1.17 Determination of water smell intensity.

Smell	Nature of smell manifestation	Intensity balls
Absent	Smell is not felt under any circumstances	0
Very weak	Smell is not noticeable to users and only discovered in careful study (when water is heated)	1
Weak	Smell is noticeable to users if to draw their attention to it	2
Noticeable	Smell is easily noticeable and makes the water unpleasant for drinking	3
Clear	Smell draws attention and forces to reject using the water	4
Very strong	Smell is so strong that it makes the water unsuitable for the use	5

of added distilled water without smell in ml. If the analyzed water contains a known odorous substance, the described method may be used for an approximate estimate of its concentration. For this purpose the threshold odor number of the standard solution of this odorous substance with known concentration N_i is first determined. Then the concentration of this substance in the sample (C_i) will be equal to:

$$C_{o,i} \approx C^*_{o,i} \frac{S^*_{o,i}}{S_{o,i}}, \tag{1.43}$$

where $C^*_{i,i}$ and $S^*_{i,i}$ are the concentration and threshold odor number of the component i in the standard solution, $C_{o,i,}$ and $S_{o,i}$ are the concentration and threshold odor number of the component i in the sample.

The smell is rapidly lost. For this reason its determination should be conducted directly in the field, at the moment of testing. Under GOST 3351-74 smell should be determined at normal (20°C) and elevated (60°C) temperature. For this reason in the field a small amount of water is collected in a tube, plugged, heated and shaken. Smell of no greater than 2 balls is accepted for drinking water.

Taste is a capacity of water to affect gustatory organs, namely taste buds of the tongue, receiving chemical signals. The taste of water is determined last of all, when the absence of danger of color and smell is ascertained. The taste of water depends on its composition, first of all on the content and nature of mineral components. For this reason the taste of natural water is used first of all for estimate of its salinity and also of admixture

Table 1.18 Determination of intensity.

Taste or after-taste	Taste and after-taste manifestation type	Intensity balls
None	Taste and after-taste are not felt	0
Very weak	Taste and after-taste are not felt by the user right away but are discovered in careful testing	1
Weak	Taste and after-taste are noticed if to pay attention	2
Noticeable	Taste and after-taste are easily noticed and cause disapproving reaction to the water quality	3
Clear	Taste and after-taste attract attention and force rejection of use	4
Very strong	Taste and after-taste are so strong that make the water unsuitable for use	5

of organic components. Four major tastes, salty, sour, sweet and bitter are distinguished. Other taste feelings are called after-tastes. After-taste may be fishy, metallic, etc. Intensity of taste and after-taste is estimated on the 5-ball scale as determined by GOST 3351-74 and listed in Table 1.18.

Taste and especially after-tastes of water may noticeably change due to the loss of volatile components and biochemical processes in the sample water. For this reason taste of the water should be determined immediately in the field. Taste Intensity is determined at 20^0 C, sometimes the water is heated to 30^0 C. Tested water is taken into the mouth in small portions and held, without swallowing, 3 to 5 seconds. For drinking water, taste and after-taste value of no greater than 2 balls is acceptable.

Balneological properties pertain to curative qualities of waters, i.e., their capacity to affect a human organism positively. These properties are caused by presence in the water of elevated concentrations of biologically active components (CO_2, H_2S, As, Ra, etc.), elevated radioactivity or temperature. Waters with clearly expressed balneological properties are called in Russia ***mineral waters***.

A joke: in a district center of Volgograd Province, Tarngord Town, geologists discovered ground water with surprising properties. It turned out that regular drinking of this water turns alcoholics in teetotalers.

Under GOST 13273-88 mineral water with salinity lower than 10 $g{\cdot}l^{-1}$, which contains ***biologically-active components*** with concentration no lower than the standard listed in Table 1.19, is attributed to drinking water. Mineral drinking waters include table waters with salinity below 1 $g{\cdot}l^{-1}$ and

Table 1.19 Balneological standards of potable mineral waters (GOST 13273-88).

Mineral water	Biologically-active component	Minimum mass content of component, mg/l
Carbonated	CO_2	500.0
Ferruginous	Fe	10.0
Arsenical	As	0.7
Boric	H_3BO_3	35.0
Siliceous	H_2SiO_3	50.0
Brominous	Br	25.0
Iodous	I	5.0
Containing organic matter	TOC.	5.0

curative-table waters with salinity no greater than 10 $g{\cdot}l^{-1}$. Mineral waters with salinity 10-15 $g{\cdot}l^{-1}$ are considered curative.

Hydrochemistry studies composition and formation conditions of mineral water whereas balneological properties, namely, their influence on our organism at their topical and ental application, are studied by physicians-balneologists.

1.6.2 Chemical Properties

Chemical properties characterize quality of water as a chemical. They determine suitability of water for usage depending on its composition. These properties include solid residue (salinity), gas-saturation, biochemical use of oxygen and oxidability, acidity-alkalinity, redox potential, hardness and corrosiveness.

Solid residue (*residue on evaporation*) describes capacity of the water to form salts at evaporation; it is measured by the mass of solid matter, which remains after the removal of the moisture from a unit of its volume. This matter is represented mostly by nonvolatile mineral components and to a smaller extent by organic matter. For this reason value of the solid residue is very close to salinity, i.e., total analytical content of mineral components.

Methods of dry residue determination are regulated by GOST 18164-72. Under this technique 250-500 ml of filtered water is evaporated over water-bath with distilled water and the remaining dry residue is dried up at 110°C until its weight is stabilized. In order to prevent hydrolysis and absorbability of Mg and Ca, prior to the evaporation soda (Na_2CO_3) or sodium fluoride (NaF) is added to the sample. The amount of the dry residue, same as the salinity, is expressed for fresh waters in mg·l^{-1}, for brakish and salt waters in g·l^{-1} and for brines often in g·kg^{-1}. In the USA the amount of dry residue is expressed in parts per million of weight (*ppm*).

The dry residue may be used for an experimental estimate of salinity value and correctness and completeness of the component mineral content analysis. At high hydrocarbon content its value may be noticeably less than the salinity due to the formation and loss of CO_2. For this reason, for fresh waters dry residue and salinity are sometimes tied together through the following equation:

$$m = m_{\text{d.r.}} + 0{,}5\left(C_{\text{w,HCO}_3} + C_{\text{w,CO}_3}\right)$$

where m and $m_{\text{d.r.}}$ are the salinity and dry residue, respectively.

On the contrary, at high salinity and high sulfate and chloride content, calcium and magnesium of the dry residue may be noticeably greater than the salinity due to the formation of crystal-hydrates, base salts and absorption of the moisture. The excess of dry residue over salinity may be also associated with high organic matter content. The weight loss after annealing of dry residue is sometimes considered as a rough estimate of the hydration water and organic matter. The difference between the values of the dry residue and total salinity usually do not exceed 3%.

Increase in dry residue and salinity deteriorates water quality, and makes it unsuitable for drinking, for agriculture and for technical use. According to hygienic requirements dry residue or salinity in potable waters must be below 1,000 mg·l^{-1}. In extreme cases, upon the approval of the Sanitary and Epidemiological Directorate, their increase to 1,500 mg·l^{-1} is accepted. Taste qualities of the underground water directly correlate with the value of dry residue (salinity), which is shown in their classification in Table 1.20.

Table 1.20 Classification of ground water by salinity.

Water category	Salinity, g·dm^{-3}
Superfresh	<0.3
Fresh	0.3–1.0
Slightly brakish	1.0–3.0
Brakish	3.0–10.0
Salty	10–35
Weak brines	35–150
Strong brines	150-300
Very strong brines	>300

Gas-saturation describes the capacity of water to release natural gas and is measured by value of its max gas-water ratio. Gas-saturation is determined directly at the moment of testing at degassing by spontaneous or forced technique. Gas-saturation at low values is usually expressed in $ml \cdot l^{-1}$ or $cm^3 \cdot l^{-1}$ and at high values in $cm^3 \cdot cm^{-3}$.

Gas-saturation is important for the estimation of balneological properties of ground water, their corrosiveness and also for reliable estimattion of their composition and properties in reservoir conditions. On the surface gas components are rapidly lost and do not have a significant effect on water quality in normal conditions. Nevertheless, fizz water may hamper some technological processes. H_2S presence may make the water unsuitable for use even at low gas-saturation. On the contrary, high CO_2 content increases the quality of mineral waters. It is conventional to consider gas-saturation below 50 $ml \cdot l^{-1}$ very low, 50 to 100 $ml \cdot l^{-1}$ low, 100 to 1,000 $ml \cdot l^{-1}$ medium, and greater than 1,000 $ml \cdot l^{-1}$ high. Ground waters usually have gas-saturation lower than 40 $ml \cdot l^{-1}$. With increasing depth, water gas-saturation may noticeably increase.

Biochemical oxygen demand - BOD describes capacity of the water to oxidize own organic matter due to the activity of own aerobic microorganisms. It is measured by the amount of O_2 necessary for oxidizing organic matter by microorganisms prior to the start of nitrification, i.e., oxidizing ammonia to NO_3. The amount of oxygen used for oxidizing ammoniacal nitrogen to nitrates and nitrites at the determination of BOD is not accounted for.

Water sample for BOD determination is collected into a glass vessel. Analysis is performed asap. Safekeeping of samples is acceptable at low temperature during a period no longer than 1 day. For BOD determination a settled sample is diluted with pure water, initial free oxygen content is determined, and then it is kept in the dark in a hermetic vessel for several days. The difference between the initial and final oxygen concentration is biochemical oxygen demand (BOD).

$$\mathrm{BOD} = C_{\mathrm{w,O,b}} - C_{\mathrm{w,O,a}}, \tag{1.44}$$

where $C_{\mathrm{w,O,b}}$ and $C_{\mathrm{w,O,a}}$ are O_2 concentrations before and after incubation, $mg \cdot l^{-1}$. BOD is expressed in milligrams of oxygen per liter of studied water ($mgO \cdot l^{-1}$).

Reliability of the BOD value depends on conditions of its measurement, composition of organic matter, content of the dissolved O_2 and duration of the incubation.

Temperature noticeably increases the rate of organic matter biochemical oxidizing, 2–3 times with the increase by 10°C. Usually the incubation is conducted at temperature 20±1°C.

The rates of biochemical oxidizing of different organic compounds are different. Monoatomic aliphatic alcohols, phenols, formaldehydes, furfurol, etc., are considered easily oxidizable (biologically soft). Slowly destroyed (biologically hard) are polyatomic phenols, non-ion-active surfactants and other more complex organic compounds. Intermediate position is taken by methylphenols, polyatomic phenols, anionoactive surfactants, etc. For complete oxidizing of a biologically stable organic matter, is required longer incubation period than for a biologically unstable one. For this reason under the same conditions and equal content a hard-oxidizable organic matter will produce lower BOD values than easily – oxidizable one.

Significant part of microorganisms has their oxygen optimum for multiplication and biochemical activity. For this reason the O_2 content affects the rate of organic matter oxidation. For achieving maximum accuracy of BOD the sample must have at least 5 $mg{\cdot}l^{-1}$ of oxygen, greater than 2 $mg{\cdot}l^{-1}$ for oxidizing and greater than 3 $mg{\cdot}l^{-1}$ in reserve. At high organic matter content, for instance in waste waters, the sample is preliminarily aerated or diluted with oxygen-saturated water with addition of adjusted microorganisms.

Finally, in water may be present nitrifying bacteria (*Nitrozomonas, Nitrobacter* and others), which cause oxidizing of nitrogen-containing compounds. They facilitate the conversion of ammoniacal nitrogen in NO_2^- and then in NO_3^-, thereby increasing dissolved oxygen demand.

Nitrification start may be recorded by a jump on the diagram of BOD daily increment during the incubation period (Figure 1.13). Nitrification begins approximately in the 7th day thereby overstating BOD value. The amount of oxygen expended for nitrification may be several times the oxygen amount for biochemical oxidizing of organic compounds. For this reason in the BOD determination for 10 or more days, it is necessary to add to the sample special substances – inhibitors suppressing life-sustaining

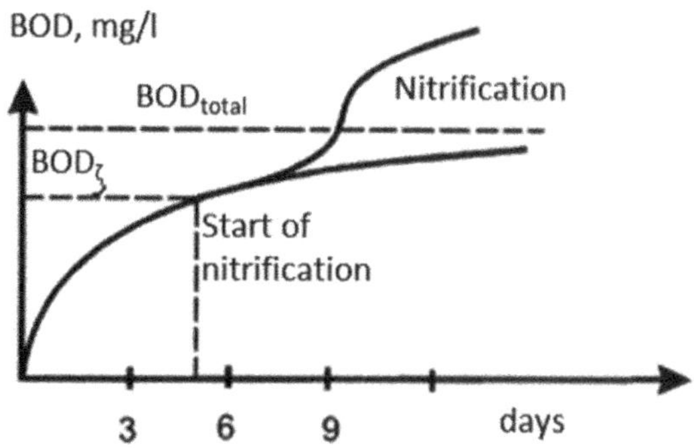

Figure 1.13 Change in oxygen demand at nitrification.

activity of nitrifying bacteria but not affecting bacteria - oxidants of organic compounds. Thiocarbamide is applied as such an inhibitor; it is added to the sample in concentrations of 0.5 $mg{\cdot}l^{-1}$.

Usually BOD is determined for 5 days of incubation (BOD_5), however, contents of some compounds are more informatively described by BOD value for 10 days or over the total oxidizing period, 20 days (BOD_{10} or BOD_{total}, respectively). It is believed that BOD_5 accounts for around 70% of BOD_{total}, which is typical for surface reservoirs polluted mostly by public-household waste waters. However, this value may range between 10 and 90% depending on the nature of organic matter.

BOD serves as a rather coarse indirect estimate of biochemically oxidizable organic matter content and is also used as a parameter of water pollution (Table 1.21). The BOD_5 value in public-drinking waters should not exceed 3 $mgO{\cdot}l^{-1}$ and in waters of public-drinking and cultural use, 6 $mgO{\cdot}l^{-1}$.

Chemical oxygen demand - COD or **oxidability** is capacity of a water solution to reduce oxygen of strong oxidants. In other words, COD describes total content of deoxidants in waters in $mgO{\cdot}l^{-1}$. Such deoxidants are mostly organic, and to a smaller extent, mineral components.

COD is measured by the amount of oxidant oxygen expended for oxidizing. There are theoretical and measured COD values. Theoretical value, COD_{theor} is maximum amount of oxygen in $mg{\cdot}l^{-1}$, necessary for total oxidizing deoxidants in the sample. It is believed that carbon is oxidized to CO_2, sulfur and phosphorus, to SO_3 and P_2O_5, nitrogen converts into ammonia and hydrogen transfers to H_2O or ammonia. Any natural water has only one value of COD_{theor}, which for many individual deoxidants may be calculated.

Measured COD depends as well on conditions of testing as on the technique of the determination. Oxidizable matter may be dissolved, suspended

Table 1.21 BOD5 values vs. pollution.

Contamination extent (types of water reservoirs)	BOD_5, $mgO{\cdot}l^{-1}$
Very clean	0.5–1.0
Clean	1.1–1.9
Moderately polluted	2.0–2.9
Polluted	3.0–3.9
Dirty	4.0–10.0
Very dirty	Over 10.0

and colloidal, which in the process of analysis is often not accounted for. For this reason oxidability of filtered samples may be noticeably lower than unfiltered. Sample water for oxidability determination is usually collected in a glass vessel. If this sample is acidified by sulfuric acid to pH below 2 and cooled to 2-5°C it may be held before analysis a maximum of 5 days. If the sample is placed in a polymer vessel and frozen to minus 20°C, it is possible to hold it for almost a month.

At the COD determination excessive amount of oxidant is introduced in the water sample. Oxidizing is performed in hard conditions by boiling in the presence of sulfuric acid. Then unused residue of oxidant is determined titrimetrically. The difference between the introduced amount and the residue of the oxidant determines COD value. Oxidability is expressed in milligrams of oxygen expended for oxidizing organic matter in 1 liter of water ($mgO_2 \cdot l^{-1}$). In the USA COD is often measured in weight parts per million (ppm).

COD value depends on the oxidant properties, water composition, duration and conditions of oxidizing.

Depending on the oxidant there are distinguished: COD-permanganate ($KMnO_4$), bichromate ($K_2Cr_2O_7$), iodate (KIO_3), etc. Measured COD is always somewhat lower than COD_{theor} but the oxidability values of the same water by different oxidants are usually non-unique due to unequal oxidizing. Most complete oxidizing is achieved when using the bichromate and iodate oxidants.

Bichromate oxidability allows obtaining values closest to COD_{theor}. This technique oxidizes most organic compounds to 95% and greater. Only toluene, benzene pyridine, paraffin and other more complex compounds practically do not oxidize. Sometimes for acceleration and increasing completeness of organic matter oxidizing, silver sulfate is added as catalyst.

Permanganate oxidability oxidizes only 25-50% of the organic matter, and noticeably understates COD values. At that, most easily oxidized is humic matter, containing much more nitrogen. For this reason permanganate oxidability is suitable for studying of the ground water with easily oxidizable humic acids. In more polluted waters bichromate oxidability is preferable.

Nevertheless, more often is determined exactly the permanganate oxidability, which is used as the measure of water contamination. This analysis is regulated by GOST 23268.12-78. Under SanPiN 2.1.4.1074-01 the extent of drinking water contamination is possible to estimate by the value of permanganate oxidability (see Table 1.22). In the waters for drinking use, the value of permanganate COD should not exceed 5 $mgO \cdot l^{-1}$, and in water of recreation zones, value no greater than 30 $mgO \cdot l^{-1}$ is accepted.

Table 1.22 Permanganate COD values as parameter of pollution.

Contamination extent	**COD, mgO·l^{-1}**
Very clean	1
Clean	2
Moderately polluted	3
Polluted	4
Dirty	5–15
Very dirty	>15

Acidity - alkalinity determines acid-alkali properties of waters, i.e., their role in exchange reactions of H^+ and OH^- ions. Many chemical and biochemical processes, form of existence of individual elements and their stability and toxicity of the polluting matter depend on these water properties.

Acidity-alkalinity depends directly on relative content in water of H^+ and OH^- ions, which is measured, as a regular rule, by the pH value and much rarer by the solution's alkalinity and acidity.

The **pH** value is a function of effective concentration of the hydrogen ion C_{H^+}:

$$pH = \log\frac{1}{C_{H^+}}. \tag{1.45}$$

The moisture dissociation constant in stable conditions is constant and equal to 10^{-14} mole^2·l^{-2}. For this reason in neutral waters where H^+ and OH^- concentrations are equal, pH value is 7. Addition of acid increases relative content of H^+ and decreases pH value below 7. Water with excessive H^+ has properties of acids. Addition of bases decreases relative content of H^+ and raises pH value above 7. At deficiency of H^+ water acquires properties of alkali.

The pH value is very sensitive to changes in water composition and environment. It may noticeably change due to the loss of CO_2, weak acids or precipitation of salts. For this reason the determination of the hydrogen parameter should be conducted directly on the spot of testing. Even a cooled sample may be safekept no longer than 6 hours.

For the pH determination is used the indicator paper, visual colorimetry or pH-meters.

Indicator paper is used for rapid pH estimate in unknown solutions. Accuracy of this technique is no better than ±1 that is insufficient for estimate drinking water qualities.

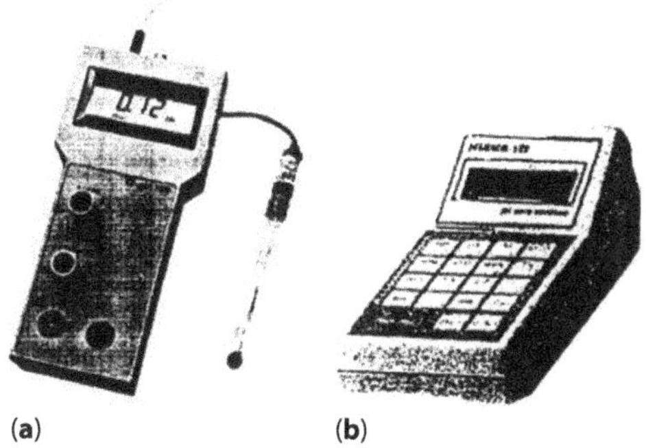

Figure 1.14 pH-meters: (a) pH-meter - millivoltmeter pH-150M intended for pH, Eh and water solution temperature determination. Power source: wall outlet or batteries. (b) **pH**-meter - ionometer "Ecotest-IO" intended for pH, Eh, ion activity (pX), COD, temperature and other water parameters and composition measurement. May work with any standard ion-selective electrodes.

Visually-colorimetric determination is conducted using portable testing kits. This technique is based on the use of reactions between the indicator and hydrogen ions; these reactions change the color of the solution. The accuracy of pH measurement with visually-colorimetric test kits is around ±0,5.

In recent times electrometric measurement methods are ever greater used, namely, pH-meters. Currently many different pH-meters exist, both of the domestic and foreign manufacturing (Figure 1.14). They measure pH value by electrodes. A glass pH-electrode is immersed in the tested solution and connected through an electric chain with another reference electrode. Reference is usually a mercury-horn mercury electrode immersed in saturated KCl solution with a permanent H^+ activity. All pH-meters require preliminarily calibration against a buffer solution. Thus, besides the device proper it is necessary to have two buffer solutions. Hydrogen parameter measurement accuracy with pH-meters reaches ±0.1 and even better.

Based on the pH value a judgment is made about water capability to participate in acid-alkali reactions. This parameter is very convenient and is commonly used for the characterization of ground water acidity–alkalinity according to the scale in Table 1.23. Under the requirements in SanPiN'a 2.1.4.1074-01, the pH value in potable waters must be within 6-9. At pH below 4 the water acquires sourish taste. Optimum for fish is the water with pH of 6.5 to 8.5, whereas water with pH below 4.4 and greater than 8.8 may cause gill irritation and even death. Waters with low pH values are also corrosive toward metals and concrete, and in combination with other factors noticeably accelerate their corrosion.

Table 1.23 Properties of ground water depending on the pH values.

Group	pH	Group	pH
High-acidity water	<3	Weak alkalinity water	7.5–8.5
Acid water	3–5	Alkali water	8.5–9.5
Low-acidity water	5–6.5	High alkalinity water	>9.5
Neutral water	6.5–7.5		

Rarely ground water's acid-alkali properties characterize its capacity to neutralize acids or alkali, i.e., alkalinity and acidity value.

Alkalinity of ground waters determines their capacity to neutralize strong acids. Its value is equal to equivalent domination of strong bases over strong acids and is determined by direct tittering of water solution by a strong acid (HCl) with lowering of pH. The methyl orange indicator is used as the indicator of pH values. The National Bureau of Standards/ National Institute of Standards of the USA recommends for the same purpose to use the mix indicators: bromine-cresol green and methyl red. The alkalinity value is measured by the amount of acid expended for neutralizing in mg-equiv·l^{-1}. In the USA alkalinity is often measured in mg $CaCO_3$ per liter of water. In order to convert these values in mg-equivalents they have to be divided by 45.

It is believed that at pH 4.5 the water has zero alkalinity. When pH is 4.2–4.5 ***total alkalinity*** is determined. The part of alkalinity, which is tittered to pH 8.0–8.2, is called ***free alkalinity***. For the determination of free alkalinity phenolphthalein is usually used as an indicator.

Acidity of ground waters describes their capacity to neutralize strong bases. Its value is equal to the equivalent domination of a strong acid over strong bases and is determined by direct water tittering by a strong alkali. The solutions of *KOH* or NaOH are usually used (concentration 0.05 or 0.1 mole·l^{-1}). The acidity value is measured by the amount of alkali expended to neutralize, in mg-equiv·l^{-1} (Figure 1.15).

Equivalent amount of the alkali necessary for raising pH to 8.3 describes ***total acidity*** of the waters. Part of the total acidity measured at tittering to pH 4.5 is called ***free acidity***. Within this range are tittered acids HCl, HNO_3, H_2SO_4 and H_3PO_4.

The water solution is electrically neutral. For this reason equivalent concentrations of components in ground water, as a regular rule, are associated through the balance of charges' equation:

$$\sum C_{\text{eq,s.b.}} + C_{\text{eq,H}^+} = \sum C_{\text{eq,s.a.}} + C_{\text{eq,OH}} + C_{\text{eq,HCO}_3^-} + C_{\text{eq,CO}_3^{2-}} + C_{\text{eq,H}_2\text{BO}_3^-} + C_{\text{eq,H}_3\text{SiO}_4^-} + C_{\text{eq,HS}^-} + \sum C_{\text{eq,o.a.}}, \quad (1.46)$$

where $\Sigma C_{eq,\,s.b.}$, $\Sigma C_{eq,s.a.}$ and $\Sigma\, C_{eq,o.a.}$ are summary concentrations of strong base cations, strong acid anions and organic acid anions, respectively, in mg-equiv·l^{-1}. This equation will stand if addend $\Sigma C_{eq,s.a.}$ is moved left and C_{eq,H^+} – right of equal mark:

$$\sum C_{eq,s.b.} - \sum C_{eq,s.a.} = C_{eq,HCO_3^-} + C_{eq,CO_3^{2-}} + C_{eq,H_2BO_3^-} + C_{eq,H_3SiO_4^-} + C_{eq,HS^-} + \sum C_{eq,o.a.} + \left(C_{eq,OH^-} - C_{eq,H^+}\right). \tag{1.47}$$

At pH around 7 the difference $C_{eq,OH^-} - C_{eq,H^+}$ may be disregarded. In most ground waters the concentrations of weak acids of boron, silicon, hydrogen sulfide and organics are negligibly small relative to the concentrations of HCO_3^- and CO_3^{2-}. For this reason the alkalinity value of these ground waters is very close to the sum $C_{eq,HCO_3^-} + C_{eq,CO_3^{2-}}$ or $(C_{M,HCO_3^-} + 2C_{M,CO_3^{2-}})$. This sum is usually called ***carbonate alkalinity***. Part of alkalinity generated by the boric acid anion is sometimes identified as ***boric alkalinity***.

Alkalinity and acidity describe not pH value but the capacity of the water to maintain this value at the addition to it of strong acids or alkali. At low alkalinity or acidity values the water is very sensitive to their action. At their high values the water serves as a buffer, i.e., stabilizer of pH values. The role of alkali buffer is usually played by the sum of HCO_3^- and CO_3^{2-}, more rarely OH^- and other weak acids. At pH below 4.5 the water practically does not contain HCO_3^- and CO_3^{2-} and does not have alkalinity, at pH below 8.3 the buffer is mostly HCO_3^- and at pH higher than 8.3 – CO_3^{2-} and HCO_3^-.

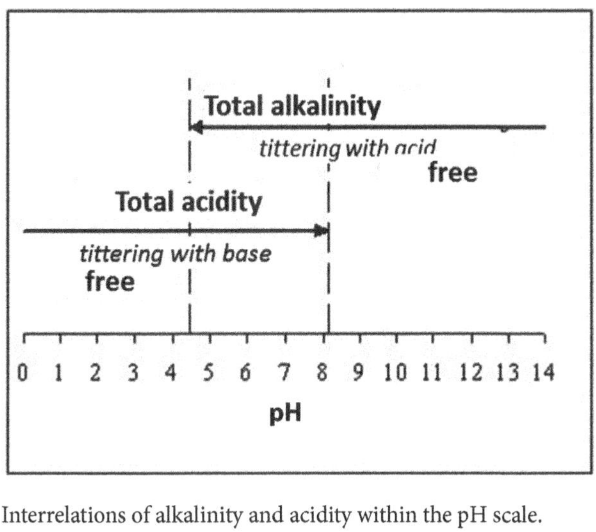

Figure 1.15 Interrelations of alkalinity and acidity within the pH scale.

Water alkalinity together with **pH** values is used to calculate the carbonate content and balance of carbonic acid in the water. Alkalinity determination with excessive concentrations of alkali-earth metals is important for estimation of water suitability for irrigation. Alkalinity determination is useful for dosing of chemicals necessary in treatment of waters for water-supply and also for the reactant cleaning of some waste waters.

Redox potential (ORP) describes the capability of natural water solution as oxidant or reductant.

The substance of redox processes is exchange of electrons by individual elements and change of their oxidation number. The solutions differ one from the other in the number of electrons potentially capable of participating in these processes. Excess of such electrons makes the solution a reductant and the deficit, an oxidant. The redox potential value is measured either directly by the concentration of these very electrons or by the value of voltage they are capable to form relative the ***standard hydrogen electrode*** (***SHE***). In the former case the redox potential is expressed by analogy with pH by a dimensionless value:

$$\mathrm{pe} = \log\frac{1}{C_\mathrm{e}}. \tag{1.48}$$

Here, C_e is concentration of electrons capable of participating in redox reactions. Value of solution's pe is impossible to measure. It is determined only by calculation from Eh value, which describes deflection of the measured solution's potential from potential of a standard hydrogen electrode accepted to be equal to 0.00 V. Eh is measured in volts or millivolts (mV). Eh and pe values are connected as follows:

$$\mathrm{pe} = \mathrm{Eh}\frac{F}{2.3RT}, \tag{1.49}$$

where F is Faraday's constant ($9.6485 \cdot 10^{-4}$ C·mole^{-1} or 23.061 kcal·V^{-1}·g-equiv^{-1}), R is gas constant ($1.987 \cdot 10^{-3}$ kcal·deg^{-1}·mole^{-1}), T is absolute temperature. Under standard conditions, i.e., at 25^{0}C:

$$\mathrm{pe} = 16.9 \cdot \mathrm{Eh} \text{ and } \mathrm{Eh} = 0.059 \cdot \mathrm{pe} \tag{1.50}$$

Eh value is determined either with electrodes under laboratory conditions or with Eh-meters in the field, or by calculation from the content of redox-pairs. Instead of a hydrogen electrode, other more reliable electrodes are used (calomel or silver chloride) whose potential is known. As a regular rule, hermetic circulation systems with inbuilt electrode and thermometer are used for Eh measurement. Measurement is preceded by flushing the device with the studied water until 3-5 volumes are replaced. The corrections depending on pH values and nature of the electrode are taken into account at calibration of the devices. It is recommended to use at least 3-4 indicator electrodes for reliable measurement.

The calculation technique is based on measuring concentrations of two components associated between themselves by oxidation-deoxidation reaction. The ratio of their concentrations allows for the estimation of the oxidation-deoxidation potential value. This technique is reviewed in more detail later.

Depending on the Eh values 3 major types of redox environment in ground waters are distinguished:

- oxidative with Eh values greater than +250 mV (pe > 4.2), O_2 content greater than 3.5 mg·l^{-1} (2.45 ml·l^{-1}) and metals of high oxidation number (Fe^{3+}, Mo^{6+}, As^{5-}, V^{5+}, U^{6+}, Sr^{4+}, Cu^{2+}, Pb^{2+});
- transitional, with Eh values from −150 to +250 mV (pe of 4.2 to −2.5), with O_2 content less than 3.5 mg·l^{-1} (2.45 ml·l^{-1}) and H_2S less than 50 mg·l^{-1} (32.5 ml·l^{-1});
- reductive with Eh values less than −150 mV (pe < −2.5) and with H_2S content greater than 50 mg·l^{-1} (32.5 ml·l^{-1}) and metals of low oxidation number (Fe^{2+}, Mn^{2+}, Mo^{4+}, V^{4+}, U^{4+}).

Ground water Eh values are in the range of +800 to −400 mV. The water with high Eh values is suitable for use. In itself, the deficiency of electrons capable of participating in oxidation-deoxidation processes is of no great danger for life. But natural water with Eh below 250 mV may contain hydrogen sulfide and other deoxidants with disagreeable smell and taste, which substantially degrade its quality.

Hardness describes the capacity of ground water to form insoluble salts under conditions of technological processes. Such salts form cations of alkali-earth (Ca^{2+}, Mg^{2+}, Ba^{2+}, Sr^{2+}) and polivalentny metals (Fe^{3+}, Al^{3+}, Mn^{2+}) with carbonates and sulphates. It is exactly Ca^{2+} and Mg^{2+} that determine most part of hardness as the rest of the metals are present usually in negligible amounts.

The following types of hardness are distinguished: total hardness, removable (temporary) and irremovable (permanent). ***Removable or temporary hardness*** is part of the total hardness, which is possible to eliminate by boiling. This hardness forms scale deposits in steam boilers, kettles and washing machines. Hardness remaining after boiling is called ***irremovable or permanent hardness***. Together removable and irremovable hardness

Table 1.24 Conversion of foreign hardness units to meq·l^{-1}.

Degrees of hardness	**meq·l^{-1}**	**Degrees**			
		German	**French**	**Englisn**	**American**
Russian	1	2.804	5.005	3.5110	50.045
German	0.3566	1	1.7848	1.2521	17.847
French	0.1998	0.5603	1	0.7015	10
English	0.2848	0.7987	1.4255	1	14.255
American	0.0200	0.0560	0.1	0.0702	1

equal to ***total hardness***. In Russian according to GOST P 52029-2003, hardness value is measured by units meq·l^{-1}, which name Russian degrees of hardness. Thus appeared Russian degrees of hardness, which line up with degrees of other countries (See Table 1.24).

Loss of CO_2 may noticeably lower the hardness value. For this reason it should be determined within a day prior to degassing. For the samples with specific electric conductivity, greater than 70 mSm·m^{-1} is acceptable safekeeping for 48 hours.

Initially hardness of water was defined as concentration of ions capable of forming insoluble deposit at interaction with the sodium (potassium) soap or with palmitic acid. Interaction between the alkali earths with fatty acids forms water-insoluble salts:

$$2Na(\text{fatty acid}) + CaSO_4 = Ca(\text{fatty acid})_2 + Na_2SO_4.$$

Currently, according to GOST 4151-72, total hardness is determined experimentally by tittering samples with the chemical trilon B (Na_2EDTA[2]) at pH 10 and in the presence of an indicator (chromium dark-blue or chromogen black). Na_2EDTA – disodium salt of the ethylenediaminetetraacetic acid, which forms firm complex salts with bivalent and trivalent metals. This trilonometric technique is applicable for waters with diverse hardness.

There are currently automatic water hardness analyzers (for instance, AKMC-1), which provide continuous automatic determination of total calcium and magnesium concentration in water (total hardness) within the range of 0.05 to 5.0 meq·l^{-1}. This technique is based on potentiometry.

It is possible to estimate the value of total hardness by calculation as total content of Ca^{2+}, Mg^{2+}, Ba^{2+} (sometimes Fe^{2+}) in meq·l^{-1} (degrees of hardness). The calculated general hardness may be noticeably lower than

[2] EDTA is Ethylenediaminetetraacetic acid.

the experimental one if the water has high content of unaccounted-for polyvalent metals.

Removable hardness is determined experimentally from the decrease in total hardness after boiling water for 1 hour. Removable hardness is caused by the presence of hydrocarbonate ions. In boiling they are destroyed and removed from the solution together with CO_2 and alkali-earth metal in the form of scales:

$$Ca(HCO_3)_2 \rightarrow \underline{CaCO_3} + H_2O + CO_2.$$

Removable hardness may be also calculated as the number of Ca^{2+} and Mg^{2+} ions equivalent to water's carbonate alkalinity. Such calculated hardness is often called ***carbonate hardness*** (*temporary hardness*). The term "*carbonate hardness*" is treated in American publications as synonym of the carbonate alkalinity. For its determination it is necessary to compare the values of total calculated hardness $C_{eq,Ca^{2+}}+C_{eq,Mg^{+}}$ and carbonate alkalinity $C_{eq,HCO_3^-} + C_{eq,CO_3^{2-}}$ in meq·l^{-1}. The carbonate hardness is equal to the lower value. Even after boiling, carbonates retain some solubility in water (especially magnesium carbonates). Moreover, their solubility increases with the increase in content of chloride and sulfate compounds. For this reason in the experimental determination of removable hardness some amount of carbonate salts remain in solution. Due to this the removable hardness, albeit close to the carbonate one, is somewhat smaller (by 1–1.5 meq·l^{-1}).

Irremovable (constant) hardness is caused by sulphates and other noncarbonate salts. It is determined as the difference between total and removable hardness values derived experimentally. Irremovable hardness is possible to estimate also from calculated values of total and carbonate hardness. In this case it is sometimes called ***noncarbonate hardness***. Calculated noncarbonate hardness may be somewhat smaller than the amount determined by experiment. Moreover, due to technical errors it may acquire negative values. In this case noncarbonate hardness is taken at 0.

Hardness noticeably affects consumer properties of waters. High hardness deteriorates organoleptic properties of water giving it bitterish taste and affecting digestive organs. Under GOST 2874-82 general hardness of drinking water should not exceed 7, and in the exceptional cases when there are no waters of a better quality, 10 meq·l^{-1}. But much more important is that hardness often makes water unsuitable for manufacturing paper, sugar, leather, etc. It facilitates scale formation on the walls of steam boilers, in washers and tea-kettles, thereby lowering their operational efficiency. Besides, for achieving equal effect in linen washing hard water requires greater amount of detergents.

Table 1.25 Ground water depending on their hardness after O.A. Alekin.

Water	Hardness, $meq{\cdot}l^{-1}$
Very soft	less than 1.5
Soft	1.5-3.0
Moderately soft	3.0-6.0
Hard	6.0-9.0
Very hard	Greater than 9.0

Depending on hardness O.A. Alekin proposed to classify natural waters as soft and hard (See Table 1.25)

Corrosiveness of the ground water is its capability to dissolve and destroy rocks and building materials by chemically affecting them. Depending on the nature of main corrosive agent in the water the following types of corrosiveness are distinguished: 1) oxygen corrosion; 2) acid corrosion; 3) alkaline corrosion; 4) carbon dioxide corrosion, 5) sulfate corrosion and 6) magnesia corrosion.

Oxygen corrosiveness is caused by the presence in the water of dissolved O_2 and occurs mostly on metallic constructions, in particular, metallic piles, water-supply, gas-supply and other pipes, cables, etc. Under combined presence of O_2 and CO_2 the oxygen corrosiveness increases.

Acid corrosiveness is associated with elevated water acidity caused by the presence of a strong acid. This corrosiveness manifested at pH less than 6.8, but its danger is especially high at pH below 5.0.

Alkaline corrosiveness is associated with an elevated content of alkaline metals in the solution, namely, $Na^+ + K^+$. Their acceptable concentration should not exceed 100 $g{\cdot}l^{-1}$. At total sodium and potassium content greater than 150 $g{\cdot}l^{-1}$ the water is corrosive for the concrete.

Carbon dioxide corrosiveness is determined by the presence in the water of corrosive CO_2, i.e., not balanced by carbonates of HCO_3^- and CO_3^{2-}. Such corrosiveness causes dissolution of calcite, which is in composition of carbonate rocks or forms protective crust on concrete, and washing out of calcium hydroxide $Ca(OH)_2$ from concrete.

The extent of this type corrosiveness depends on temperature and water composition and is determined by calculation or experiment. Calculated data are applicable to slightly saline waters and give only a general idea of corrosiveness. A better technique for the determination of water corrosiveness toward concrete is experimental. It is based upon the determination of water alkalinity before and after its interaction with crushed calcium carbonate. The results show the content of corrosive carbonic acid and the amount of calcium carbonate dissolved in 1 liter of studied waters.

The standards and amendments developed at VNIIG in the name of B. E. Vedeneyev of the USSR Ministry of Energy to a new draft GOST (GOST 4796-49) envision different acceptable corrosive carbonic acid content depending on HCO_3^- and total salinity values and also conditions, in which the corrosion occurs (construction thickness, filtration factor, water-head, cement grade). Depending on the content of cement and conditions, under which the construction is positioned, the water, under the new standards, has leaching corrosiveness at minimum HCO_3^- content of 0.4 to 1.5 meq·l^{-1}. Maximum corrosive CO_2 content acceptable under most dangerous conditions is 3 g·m^{-3} (1.5 ml·l^{-1}) and under least dangerous, 8.3 g·m^{-3} (4.2 ml·l^{-1}). If the cement is in contact with waters containing greater than 20 g·m^{-3} (10 ml·l^{-1}) of CO_2, it is recommended to coat it with a more resistant material.

Sulfate corrosiveness occurs at high SO_4^{2-} content. Penetrating the concrete, such water facilitates the formation of gypsum ($CaSO_4 \cdot 2H_2O$), calcium sulfoaluminate (Deval Salt), etc. This results in the volume expansion and also concrete loosening. Exactly the rapid destruction of such constructions in Western Canada early in the XX century was the cause of a broad program of water corrosiveness study. The result was special cement of elevated strength toward sulphates. Sulfate corrosiveness depends on conditions of the construction and the content of the chlorine ions. Most suitable are sulfate-resistant cement or cinder concrete. With regular cements danger occurs at SO_4^{2-} content greater than 250 g·m^{-3}, with sulfate-resistant cements - greater than 4,000 g·m^{-3}.

Magnesia corrosiveness occurs at high Mg^{2+} content. At Mg^{2+} content less than 300 mg·l^{-1} the medium is considered noncorrosive. Max acceptable magnesium content depends on cement grade, conditions, construction design and varies between 500 and 750 mg·l^{-1} and greater. Sulfate and magnesia corrosiveness is established under standards or experimentally, from change in concrete sample strength in certain water.

Corrosiveness often has complex nature and depends on content of several components. For this reason, as a regular rule, it is estimated with respect to specific materials. For instance, water corrosiveness to concrete in temperature range 0 to 25°C may be determined from Table 1.26 borrowed from SNiP 2.03.02-86 "Concrete and reinforced concrete constructions from compact silicate concrete." The medium is considered weakly-corrosive if the depth of concrete destruction over 50 years does not exceed 2 cm.

Another example is ground and other water corrosiveness to cables. Under GOST 9.602–89 "Unified system of protection against corrosion

Table 1.26 Water corrosiveness to concrete in temperature interval 0 to 25°C (SNiP 2.03.02-86).

Indications of liquid media corrosiveness	**Extent of media corrosive effect on constructions**		
	noncorrosive	**Weakly-corrosive**	**Mid- and strongly-corrosive**
1. Total corrosiveness, hydrogen parameter pH	> 4	1 – 4	< 1
2. Content of magnesium salts (magnesia corrosiveness), concentration of Mg^{2+} ions, $mg{\cdot}l^{-1}$	< 300	300 - 500	> 500
3. Content of free (corrosive) carbonic acid (carbon dioxide corrosiveness) CO_2 , $mg{\cdot}l^{-1}$	< 20	20 - 50	> 50
4. Alkaline corrosiveness, concentration caustic alkali recalculated for $Na^+ + K^+$, $г{\cdot}l^{-1}$	< 100	100 - 150	> 150
5. Content of sulfates (sulfate corrosiveness) converted to SO_4^{2-} ions, $г{\cdot}l^{-1}$	Up to 10	От 10 до 20	Св. 20

and ageing. Underground constructions," this corrosiveness depends on the nature of cable winding. For lead winding too acid (pH < 6.5) or too alkaline (pH >7.5) waters are dangerous and also elevated content of organic matter (>40 $g{\cdot}m^{-3}$) and nitrates (>40 $g{\cdot}m^{-3}$); for aluminum winding elevated content of chlorine and iron ions is dangerous (Table 1.27).

First standard documents taking water corrosiveness into account are dated back to 1949 when GOST 4796–49 "Concrete hydrotechnical. Indications and standards of water-medium corrosiveness" was proposed as the Union's standard. Technical progress in the development of corrosion protection resulted in nomenclature and material properties changes. GOST of 1949 became outdated and lost validity. Currently there are no

Table 1.27 Corrosiveness of ground and other waters to cable aluminum winding.

Corrosiveness of ground and other water	pH	Mass fraction of components, $mg \cdot l^{-1}$	
		Chlorine ion	Iron ion
Low	6.0 to 7.5	Up to 5.0	Up to 1.0
Medium	4.5 to 6.0 7.5 to 8.5	5.0 to 50	1.0 to 10
High	< 4.5 and > 8.5	> 50	> 10

unified parameters of quality for anticorrosion coating of a concrete system. The basic standards are currently under GOST 9.602–89 "Unified system of protection against corrosion and ageing. Underground constructions," SNiP 2.03.11–85 "Protection of building constructions against corrosion" and SNiP 2.03.02-86 "Concrete and reinforced concrete constructions from compact silicate concrete."

1.6.3 Physical Properties

Physical properties describe properties pertaining to natural water as a liquid medium, which does not have its own shape but almost does not change volume. Most important among them for the determination of the ground water status in reservoir are pressure and temperature, whose values regularly increase with depth; compressibility, density, electric conductivity, heat capacity and radioactivity.

Pressure describes the intensity of interaction between the water and the outside medium. It serves as one of the major parameters in its thermodynamic state, is measured by the value of force related to a unit of its area and is often called ***formation pressure***. Formation pressure is measured in pascals ($1MPa=1 \cdot 10^6 H \cdot m^{-2}$=10.19 at = 9.87 atm = 10 бар), technical atmospheres (1 at = $1kgf \cdot cm^{-2}$ = 98.0665 kPa = 0.980665 atm), physical (1 atm = 760 mm mercury = 101.325 kPa) or in bars (1 bar = $1 \cdot 10^5$Pa).

Formation pressure is a property, which is impossible to preserve in an open sample. It must be determined directly in place of underground water location. At shallow depths, in surface and ground waters, this pressure differs little from atmospheric and is tentatively assumed equal to 0.1 MPa. But with increasing depth the value of this pressure grows. Currently depths are reached where formation pressure is exceeding 100 MPa.

Figure 1.16 Downhole remote manometer AMT-08 (diameter 32 mm, length 1,010 mm) intended for measurement and recording of pressure and temperature values along the wellbore and (or) their changes in time at any point, for instance, at bottomhole for plotting a pressure reduction curve. The upper cutoff of the excess pressure measurement is 100 MPa, temperature measurement range: –20 to +125°C.

Formation pressure at great depths is either measured by special downhole manometers (Figure 1.16) or estimated by way of calculation from the values of static pressure at wellhead and weight of the water column in borehole. Requirements to manometers are determined under GOST 15807–93. These manometers may be autonomous, recording parameters on a diagram, blank paper or magnetic carrier and remote, transferring parameters to the surface through a cable.

Such physical properties of underground waters as density, compressibility, viscosity, etc., depend on the values of formation pressure. But the greatest influence formation pressure exerts is on the solubility of volatile, first of all gas, components. For this reason with growth of pressure, the content of these components may increase, and therefore, ground water gas-saturation may increase.

Used by Russian Antarctic expedition at the "Vostok" station at well depth greater than 3.5 km.

Temperature is average kinetic energy of disorderliness in the motion of components within the waters at the state of thermodynamic equilibrium. As P.L. Kapitsa wrote, "...the measure of temperatures is not the motion itself but chaoticity of this motion." Temperature is one of most important of underground water's thermodynamic state parameters, and is measured in degrees of Celsius (°C) or Kelvin (K). Great Britain and North America use degrees Fahrenheit (°F).

$$T_C = 273.16\ \mathrm{K} = 5(T_F - 32)/9, \tag{1.51}$$

where T_C is temperature in degrees Celsius, T_F – temperature in degrees Fahrenheit.

Temperature (like pressure) noticeably and rapidly changes at testing. For this reason it is measured directly at the spot. For this purpose a thermometer is run directly into the water source (shaft well or drilling well). Temperature measuring in shallow wells is done by special thermometers set into metallic casing. Maximum or lazy thermometers are used. They keep measured temperature at their extraction on the surface. For measuring temperatures at great depths electric thermometers are applied (Figure 1.16).

Accurately and timely measured temperature contains important information about conditions of water existence in situ. Temperature range of ground water is very broad, from minus 5^0 C and lower to $+150^0$ C and higher. For shallow waters ground water temperature changes during a year. With depth, water temperature loses its seasonal fluctuations and reaches stable value, usually equal to average temperature on the surface. Deeper yet, water temperature continuously increases as described by the following equation:

$$\mathrm{T}_{C,\mathrm{res.}} = \mathrm{T}_{C,\mathrm{O}} + 0.01(H - h)\frac{dT}{dH}, \tag{1.52}$$

where $T_{C,0}$ is average annual temperature at the surface (or of the neutral layer); H is depth to the studied waters; h is depth to the neutral layer; dT/dH is geothermal gradient. The value of this gradient is usually measured at a depth greater than 100 m and is, as a regular rule, in the range of 0.5 to 20^0C/100 m, but most commonly close to 3.3°C/100 m.

Temperature facilitates growth in the solubility of mineral components. For this reason at great depths the underground water is capable of having much greater salinity than near the surface.

Ground water pressure and temperature noticeably affect their density, compressibility, viscosity and other physical properties. They determine the phase state (solid ice, liquid or gas) and parameters of critical state of a water solution. However, these critical parameters of water state also depend on water composition.

Mineral components, as a regular rule, lower freezing temperature, whereas gas components lower boiling temperature. According to Robert

J. Bodnar and John K. Costain (1991), physical properties and also phase state of water brines in reservoir conditions with sufficient reliability may be modeled by the H_2O–CO_2–NaCl system (Bodnar and Costain, 1991). Addition of NaCl facilitates increase in critical temperature and pressure, but the addition of CO_2 lowers critical temperature and heightens pressure. Such critical offsets, due to the presence of dissolved salts, mostly NaCI, and gas components, mostly CO_2, may have a great significance for understanding of geologic processes associated with heat transfer by a mass of ground water. Most optimal conditions for heat transfer have brines containing around 200 $g{\cdot}kg^{-1}$ of NaCl at the temperature interval 560–620°C and pressures 700–1,100 bar, or fresh water at the temperature around 400°C and pressure 300 bar (Bodnar and Costain, 1991, pg. 985).

Compressibility serves as measure of reversible ground water compaction under the external pressure. It is described by the ***compressibility factor***:

$$\beta_w = -(dV/dP)/V_w, \tag{1.53}$$

where β_w is water compressibility factor, Pa^{-1}; V_w is water volume, m^3; P is pressure, Pa. Therefore:

$$\Delta V = -\beta_w \cdot \Delta P \cdot V_w. \tag{1.54}$$

Water, as other liquids, has negligible compressibility. Its value in fresh water at 20°C is equal to 0.5 GPa^{-1}. In order to lower the volume of water by 1% it is necessary to increase pressure by 20 MPa, which may occur at depths greater than 2 km. For this reason for shallow depths water compressibility is disregarded. Only at great subsurface depth pressure may noticeably affect the volume of water, which needs to be taken into consideration. But with depth temperature also increases, which facilitates expansion of the water. Thus, the volume of each mole of water decreases with the growth of pressure but increases with the growth of temperature. As a result, compressibility of fresh water, for instance, at pressure of 41 MPa and temperature 93° is equal to just 0.4 GPa^{-1}.

Compressibility of ground water noticeably depends on the water composition. The influence of mineral components is ambiguous. Some of them increase, and some others lower its value. A very important factor in this case is the relative content of ions with different charge density. Gas components definitely decompact solutions. For this reason, the more gas-saturated the ground water, the higher its compressibility. This correlation may be described by an empirical equation (D.L. Katz et al., 1965):

$$\beta_{w.} = \beta_{H2O}(1 - 0{,}0495Q_g) \tag{1.55}$$

where β_{H2O} is distilled water compressibility, Q_g is gas-water ratio, $m^3{\cdot}m^{-3}$. Due to this with lowering of formation pressure the volume of gas-saturated water increases insignificantly until first, the dispersed gas bubbles are released. In the further lowering of formation pressure the compressibility will be dependent on the volume of these gas bubbles.

Density of underground water is the mass of its unit volume. The denomination is $g{\cdot}ml^{-1}$, $kg{\cdot}l^{-1}$ or $kg{\cdot}m^{-3}$. Density should not be confused with ***specific gravity***, which is the force of gravity pull of mass per unit volume. Whereas the weight is determined relative to the reference weight, i.e., in gf (G) or kgf (kG), the specific gravity numerically is equal to density. In some European countries kilogram-force is officially called kilopond (*kp*). Genuine specific gravity is expressed in newtons per cubic meter ($N{\cdot}m^{-3}$). In this case:

$$\gamma_w = 9.80665\rho_{w,} \tag{1.56}$$

where γ_w is specific gravity in $N{\cdot}l^{-1}$, ρ_w is density in $kg{\cdot}l^{-1}$, 9.80665 is free fall acceleration in $m{\cdot}s^{-2}$.

Density and specific gravity values may be estimated either directly weighting certain volume of water in pycnometer or with an areometer.

These ground water properties depend on temperature, pressure and composition. At one and the same composition the greatest density (specific gravity) is observed at temperature 4°C. With temperature deflection, water density decreases from this value. Pressure compresses water and increases its density. For this reason density of underground water of similar composition noticeably decreases in depth with increase in temperature, and slightly increases with increase in pressure. At the depth of around 4-5 km the increase in underground water density or specific gravity due to changes only in temperature and pressure is just 1.6%.

However, water density and specific gravity depend much more on its composition. With the increase in content of mineral components, water specific gravity noticeably increases (Table 1.28). The increase in salinity by 1 g results in the growth in specific gravity of the solution a little less than by 0.1%. At the salinity less than 5–10 $g{\cdot}l^{-1}$ the tie between specific gravity and salinity depends little on composition of the mineral components. For this reason for fresh and brakish water, approximately to 7 $g{\cdot}l^{-1}$, ground water specific gravity usually is assumed to be 1 $g{\cdot}ml^{-1}$. Specific gravity of the sea water is 1.02-1.03 $kg{\cdot}l^{-1}$. Specific gravity reaches 1.07 $kg{\cdot}l^{-1}$ at the salinity around 100 $g{\cdot}l^{-1}$ and at the salinity around 300 $g{\cdot}l^{-1}$– 1.2-1.3 $kg{\cdot}l^{-1}$. This allows evaluating the water salinity by its specific gravity and vice versa, specific gravity (density) by the salinity, either from the Table (Table 1.28) or from the empiric equation:

$$\rho_w=1+6.3{\cdot}10^{-7}m^2+6.8{\cdot}10^{-4}m, \tag{1.57}$$

Table 1.28 Correlation of water specific gravity vs. water salinity at 15°C.

Specific gravity $g \cdot ml^{-1}$	Salinity, $g \cdot kg^{-1}$	Specific gravity $g \cdot ml^{-1}$	Salinity, $g \cdot kg^{-1}$	Specific gravity $g \cdot ml^{-1}$	Salinity, $g \cdot kg^{-1}$
1.001	2	1.037	52	1.091	120
1.002	4	1.04	55	1.095	125
1.003	6	1.042	57	1.099	130
1.005	8	1.043	60	1.107	140
1.007	10	1.046	62	1.116	150
1.008	12	1.048	64	1.125	160
1.01	15	1.049	66	1.134	170
1.012	17	1.051	70	1.143	180
1.014	20	1.053	72	1.152	190
1.016	22	1.055	74	1.161	200
1.018	25	1.056	76	1.17	210
1.02	27	1.058	78	1.18	220
1.021	30	1.059	80	1.19	230
1.023	32	1.063	85	1.2	240
1.025	35	1.067	90	1.21	250
1.027	37	1.071	95	1.22	260
1.029	40	1.074	100	1.23	270
1.03	42	1.079	105	1.241	280
1.032	45	1.083	110	1.252	290
1.033	47	1.087	115	1.262	300
1.036	50				

where ρ_w is water specific gravity in $g \cdot ml^{-1}$, m is salinity in $g \cdot kg^{-1}$. Water specific gravity may be used for a rough estimate of its salinity:

$$m = 3{,}381.3\,\rho_w - 991.0\,\rho_w^{\,2} - 2{,}388.9. \qquad (1.58)$$

Density and specific gravity of the water are necessary for the conversion of solution volume units to weight and the other way around. Exact water density determines the nature of its convection migration and tempo of hydrostatic pressure increase with depth.

Viscosity describes the strength of water's internal friction. This strength is measured in units of ***dynamic viscosity*** and ***fluidity***. The unit of the dynamic viscosity in SI system is $Pa \cdot s = N \cdot s \cdot m^{-2} = kg \cdot m^{-1} \cdot s^{-1}$. The previously used unit was balanced and equal to 0.1 Pa·s. Fluidity is the inverse value of dynamic viscosity:

$$\varphi_w = \frac{1}{\eta_w}, \tag{1.59}$$

where ϕ_w is fluidity and η_w is dynamic viscosity. Fluidity is measured in $Pa^{-1} \cdot s^{-1}$. Viscosity value is measured by viscometers.

Viscosity of ground water depends on temperature, salinity and composition. With the increase in temperatures viscosity of fresh water decreases from $1.31 \cdot 10^{-3}$ Pa·s at 10⁰C to $1.0 \cdot 10^{-3}$ Pa·s at 20° and to $0.28 \cdot 10^{-3}$ Pa·s at 100°C. The empirical correlation viscosity vs. temperature may be expressed by an equation (Touloukian et al., 1975):

$$\eta_w = 2.4 \cdot 10^{-5} \cdot 10^{\left[248/\left(T_C + 133\right)\right]}, \tag{1.60}$$

where η_w is viscosity in $kg \cdot m^{-1} \cdot s^{-1}$and T_C is temperature in °C. Because of this viscosity decreases 10 times as temperature increases from 0 to 150°C. At an average geothermal gradient of $25°C \cdot km^{-1}$ such temperature corresponds with the depths of 5-6 km.

The salinity also noticeably affects viscosity. Fresh and low-salinity water at 0°C and 101 kPa has viscosity close to $1.8 \cdot 10^{-3}$ Pa·s. With the salinity increase to 80 $g \cdot l^{-1}$ viscosity increases almost proportionally, at a greater salinity the tempo of viscosity growth increases. The presence of magnesium and calcium chlorides in brines facilitates viscosity growth. Nevertheless, the effect of temperature is usually much stronger. In connection with this is often used a parameter of kinematic viscosity:

$$\nu_w = \frac{\eta_w}{\rho_w}. \tag{1.61}$$

Viscosity strongly affects ground water mobility. The higher the viscosity, the lower the flow and mass transfer rate within the same rocks.

Electric conductivity is the capacity of ground water to let electricity pass through it. It is inversely proportionate to ***specific resistance***, i.e., electric resistivity of water between two parallel electrodes, 1 m^2 each, placed at a distance 1 m from one another. Specific electric conductivity is measured in siemens ($1S = ohm^{-1}$). Thus, its dimension is $S \cdot m^{-1}$ and that of specific resistivity is ohmm. The USA often uses *hmo* equal to 1 $S \cdot cm^{-1}$ as the units of specific resistivity.

Specific electric conductivity does not depend on low or moderate turbidity and may be determined in an unprocessed as well as in a filtered

Figure 1.17 Pocket conductometer-salinometer DIST 2 with automatic temperature compensation and graphite measurement cell.

sample. For measuring specific electric conductivity are used ***conductometers*** or ***salinometers*** (Figure 1.17).

Measurement range 0 to 10,000 $mg \cdot l^{-1}$, accuracy ±2% of the total scale, hand calibration.

Electric conductivity of natural water depends mostly on concentrations of the dissolved mineral salts and temperature.

In diluted solutions of strong and weak electrolytes the value of electric conductivity increases due to an increase in the number of ions so that it is proportionate to their concentrations. For this reason at specific conductivity below 1 $S \cdot m^1$, the correlation between its value and ion concentration is often assumed linear. For most mixed type ground waters the specific electric conductivity in $\mu S \cdot m^{-1}$ is connected with the salinity value as follows:

$$\sigma_w = \frac{1}{\omega_w} = \frac{m}{0.65}, \tag{1.62}$$

where σ_w is specific electric conductivity, $\mu S \cdot m^{-1}$, m is the salinity value, $mg \cdot l^{-1}$, ω_w is specific resistivity in $\mu ohm \cdot cm$. The multiplier 0.65±0.11 is acceptable only for relatively diluted solutions and noticeably increases when general salinity exceeds 2,000 – 3,000 mg·m^{-1}. For particular water types, there may be derived more precise correlation. For the water with noticeable concentrations of free acids, caustic alkali or sodium chloride, this multiplier may be much smaller than 0.65. In concentrated solutions due to the increase in their viscosity, ion motion velocity decreases, but together with that electric conductivity declines. Besides, in weak electrolytes the increase in salinity noticeably declines their dissociation, therefore, number of ions carrying electricity. Nevertheless, the values of ground water's electric conductivity make it is possible to judge their approximate salinity.

With an increase in the temperatures of solutions the specific electric conductivity grows. The conductivity increases approximatelyo by 2% per 1°C, and the viscosity temperature factor changes almost in proportion.

The value of the specific electric conductivity of distilled water is 3.85 μS·m^{-1}. Specific electric conductivity of the atmospheric precipitates with salinity of 3 to 60 mg·l^{-1} is within the range of 20–120 μS·m^{-1}. In fresh waters it is 30 to 3,000 μS·m^{-1} and increases to 5,000 -1,200,000 μS·m^{-1} in brines. In underground water the electric conductivity is 1 to 50 S·m^{-1} and in sea water – 3.3 S·m^{-1}. For this reason all natural water is attributed to good conductors.

The direct correlation between electric conductivity of ground water and its salinity is used in halometers (salt gauges), devices for the determination of total water salinity. This property is very important for electric survey techniques. It is exactly the underground water that is the major electro-conducting medium in the subsurface, and these properties are due to dissolved electrolytes.

Radioactivity of ground waters is their capability to radiate α, β particles and gamma-rays of the dissolved unstable (radioactive) isotopes. Thus, distinguished are alpha-activity, beta-activity and gamma-activity.

α- activity is a stream of heavy positively charged particles from protons and neutrons (helium nucleus). This stream has very low penetrating capacity and is incapable of getting through the external skin layer of organisms.

β- activity is a stream of electrons, which possesses somewhat greater penetrating capacity and may enter organisms' tissue for 1–2 cm.

γ- activity is electromagnetic radiation with wave length less than 10^{-8} cm. Gamma-radiation has huge penetrating capacity and may be blocked only by a thick lead or concrete slab.

Major radioactivity sources in the water are potassium-40, rhubidium-87, uranium-238, thorium-232 and products of their decay. A smaller role belongs to radioactive iodine, plutonium, strontium and cesium. In the final analysis most radioactivity in the ground water is associated with dissolved radon.

Radioactivity in the waters is measured by ***specific radioactivity***, i.e., by number of decays per unit of water volume per second. One decay per second is 1 becquerel. So the specific radioactivity is measured in $Bq{\cdot}m^{-3}$. Previously specific radioactivity was measured in emanation units (1 Eman = 10^{-10} $Ci{\cdot}l^{-1}$ = 3,700 $Bq{\cdot}m^{-3}$) and curies (1 Ci = $3.7{\cdot}10^{10}$Bq = 37GBq).

Radioactivity is determined either directly by rediometers graded based on standard solutions of radio-nucleids or is calculated from the content of dissolved radio-nucleids.

A water sample for radioactivity may be collected in any clean vessel. If there is a need of a separate radioactivity determination of the dissolved and suspended matter, the sample is immediately filtered. Duration of sample safekeeping without conservation should not exceed 4 hours. If nitric acid is added to 1 liter of a sample to bring it to a pH below 1, and the sample is held in the dark at a temperature 0°C – 8°C, the safekeeping may be up to 4 days. The determination methods of total specific alpha-activity of radio-nucleids in drinking water are regulated under GOST P 51730-2001.

Moderate dosage of radioactivity may be beneficial but in excess it is a great danger for health. Ground water classification by their radioactivity is included in Table 1.29.

Radioactivity is dangerous as it damages the cells of living organisms. This damage is all the more greater as more energy is transmitted to organism tissues. The amount of energy transmitted to an organism is called *dose*. The amount of radiation energy absorbed by a unit of mass of the irradiated body is called *radiation dose* or *absorbed dose* and is measured in SI system in grays (Gy) or in rads. One gray is equal to one joule per kilogram or 100 rads.

Table 1.29 Ground water classification by their radioactivity.

After E.S. Burkser (1887–1963)	After A.M. Ovchinnikov for mineral water	Radioactivity, $Bq{\cdot}m^{-3}$
Very weakly radioactive	Weakly radon	$<10^4$
Weakly radioactive		$10^4 - 10^5$
Radioactive	Medium radon	$10^5 - 10^6$
Strongly radioactive		$10^6 - 10^7$
Very strongly radioactive	Strongly radon	$> 10^7$

Ground waters possess the highest radioactivity in uranium deposit oxidizing zones. The highest recorded water specific radioactivity in the water-supply system reaches 10^8 $Bq \cdot m^{-3}$. Under SanPiN 2.1.4.559–96, accepted in drinking water is a total α-radioactivity no greater than 100 $Bq \cdot m^{-3}$ and total β-radioactivity no greater than 1,000 $Bq \cdot m^{-3}$. If total activity standards are exceeded, the identification of the water's radio-nucleids and measurement of their individual concentrations are conducted.

2

Hydrogeochemical Testing

Hydrogeochemical testing encompasses the entire complex of measures involved in the formation of data base of ground water properties and composition within the set boundaries of the studied geologic object. These measures include both collection of the ready material and measuring parameters, as well as properties and composition directly in the field or in special laboratories.

Ground waters, as all natural waters, look almost identical. In rare cases they may be distinguished by their organoleptic properties. In hydrogeochemistry they are discriminated by a number of values pertaining to their composition, chemical and physical properties. Measurement of these values is the major objective of hydrogeochemical testing.

Hydrogeochemical testing is conducted under special conditions, and that determines its specifics. First, it is associated with geologic objects of often huge size with limited accessibility to ground water. Study objects may reach hundreds of square kilometers; ground waters may be embedded at the depth of several kilometers. Second, ground water composition and properties are not stable in time. Therefore, measurements of its parameters may be dependent on the season and climatic environment. Third, ground waters are complex multi-component solutions whose understanding depends on the number and nature of studied components and techniques of their analysis. Fourth, underground water is very sensitive to changes in the situation and changes its properties and composition in the process of testing. Thus, it is necessary to have a specific idea about the acceptable error in the value of measured parameters.

In its entirety the reviewed hydrogeochemical testing is a very complex, expensive and responsible action, which may involve drilling, field, analytical and office works. It requires careful preliminarily logistical preparation, which may be subdivided into four stages: 1. Formulating the assignment and purpose of study, 2. Setting up field operations, 3. Conducting field testing, 4. Measuring the water property parameters and composition.

2.1 Assignment and Purpose of Hydrogeochemical Testing

Planning the hydrogeochemical testing depends, first of all, on the nature of issues, for which a solution is being set up. Such issues may be:

- Hydrogeological survey and monitoring observations for the purpose of providing the country's economy with necessary data about the regional natural conditions of a country's territory;
- Hydrochemical search of economic deposits, in particular, potable or mineral water;
- Evaluation and sanitary quality control of domestic/potable value ground water;
- Monitoring changes in the composition of ground water recovered at the exploitation of economic deposits;
- Finding and monitoring the distribution of dangerous components in ground waters;
- Expert evaluation of the risk, causes of contamination and ecological liability of legal entities or private individuals.

No matter how diverse the set issues are, they all require for their solution a representative base of reliable data of ground water properties and composition. That may be achieved only through hydrogeochemical testing. Issues and objectives determine only conditions under which the hydrochemical testing is expected to be conducted, namely, its regime, type and size of the studied object and also the set of measured parameters and acceptable error of their measurements.

2.1.1 Regime and Scope of Testing

Regime of testing pertains to the order of studying ground water composition and properties in time. There are no waters with absolutely permanent quality parameters in the nature. Nevertheless, in solving some issues these changes may be disregarded, whereas in solving some others they must be accounted for. Depending on the assignment of all hydrogeochemical objects, it is possible to tentatively subdivide them into stationary (stable) and nonstationary, and the testing regimes, into ***single-time*** and ***monitoring*** or ***regime*** (*repeated*).

In the first case the testing is conducted only once in a specifically assigned limited time interval, and changes in water properties and composition during this interval are disregarded. An example may be sanitary evaluation of drinking water qualities of a shaft well, drilling well or water intake, a hydrogeological survey, the determination of pollutant dispersion halo in ground waters, etc. In this case it is very important to correctly select the moment for testing and perform it within as short a time period as possible in order to minimize the influence of time.

In the second case, on the contrary, for the solution of the assigned tasks, it is required to determine changes in water properties and composition in time. For this purpose it is necessary to conduct monitoring observations with repeated testing of the same water points within a day, year or several years. For the solution of these issues, it is important to correctly plan the regime testing/monitoring and to select or prepare special observation points or wells.

The type of studied object is determined by its size and areal extent. The object may be underground water of an individual shaft well, of a cluster of water-intake wells, of a water horizon, of an individual complex or of the entire artesian basin. All these objects are first of all distinguished by the size. However, studying some of them the size is disregarded, and in others water properties and composition changes in depth are not taken into account. In connection with this, by the size and spatial dimensions it is possible to identify three basic types of hydrogeological testing objects: local, regional (two-dimensional) and volumetric (three-dimensional; 3D).

Local objects are individual water points (shaft well, drilling wells, drilling well clusters, etc.) or small portions of water-saturated reservoirs, where only average property and composition parameters are required for the description of the water. In the case of a local object, changes of the water composition and properties in 3D are not studied.

Regional objects are those of substantial areal development with specific boundaries and of limited thickness. Examples may be ground water or individual isolated underground water reservoirs, horizons in hydrogeological survey. Of interest in this case are regional changes in water composition and properties or contents of individual components. Their changes in depth are disregarded. Each water point over the area of such object is a single testing location. The final result in the description of such object is the hydrogeochemical map.

Volumetric, 3D objects are those requiring the description of water properties and composition changes both over the area and in depth. Such objects may include several water-saturated horizons separated by aquicludes (water barriers, seals). Any water point in their territory may have several testing points in depth. Drilling and completion of special wells is often required for studying volumetric objects. For the description of such objects, besides hydrochemical maps, hydrogeochemical cross-sections (profiles) are commonly used.

In the final analysis, all kinds of tasks for hydrogeochemical testing may be classified depending on the type and regime of studied objects (Table 2.1).

2.1.2 Measured Parameters and Their Errors

Selection of measured parameters and their acceptable error also depends on the assigned issue and, in its turn, determines the analytical base of

Table 2.1 Main types of issues at hydrogeochemical testing.

Type of object	Regime testing/monitoring	
	Single-time	**Monitoring**
Local	*Evaluation of water qualities of individual shaft wells, wells or cluster wells*	*Monitoring changes in water qualities in individual shaft well, well or well cluster*
Regional	*Hydrogeochemical mapping, hydrochemical exploration methods*	*Monitoring of regional contamination or moving of boundaries between fresh and salt waters*
Volumetric (3D)	*Determination of contamination boundaries, fields of potable or mineral ground water, etc.*	*Monitoring of contamination or changes in water composition with depth, discharge of salt waters, etc.*

hydrogeochemical testing, namely, the selection of necessary techniques and means of their measurement.

Measured parameters, as a regular rule, include organoleptic, some physicochemical properties (salinity, gas-saturation, pH, temperature, density, electric conductivity, etc.) and composition determined in general analysis and in special studies.

General analysis, which gives us the idea of macro component content and of major properties of the studied water, is conducted practically in any hydrochemical studies. General analysis, abbreviated (special) or complete, gives us an idea of ground water hydrochemical type, allowing us to differentiate and compare them by the composition and make first conclusions regarding their formation conditions.

Special studies depend directly on the nature of the issue/task and may include analyses of individual mineral content, gas or organic components, microorganisms, Eh, radioactivity, element isotope composition and other physicochemical parameters. For instance, in the case of drinking water quality estimate, the analyses of micro components (Fe, Mn, Cu, Zn etc.), complex organic compounds and bacteriological contamination are necessary. In the estimation of water qualities for technical use it is necessary to determine hardness and the content of Fe, Al and CO_2. Significant attention in ecological studies is devoted to technogenic pollutants and bacteriological composition. Ever greater significance acquire the studies of heavy metals, petroleum products, and pesticide contents entering together with industrial and agricultural discharge. In solving the issues of ground water composition, origin and formation, great significance belongs to halogens (Br, I, F), Ba, Li, B. Petroleum hydrogeologists devote special attention to the contents of gas and organic components in the formation water.

Required reliability of measured parameters depends on assigned tasks and liability for their solution. Under liability are understood consequences of errors ("cost of errors"). The higher the liability, the more accurately properties and composition parameter values must be measured.

Requirements to the reliability pertain to numerical parameters measured mostly in special studies, such as contents of micro components, gas or organic compounds, microorganisms and the isotope composition. The measure of reliability is ***error***, which is determined by the deviation of derived values from their sought-for true values.

The nature of sought-for values strongly depends on the objectives of a hydrochemical study. In studies of underground waters as raw material, for instance for domestic/potable use, the reliability of water properties and composition under normal (technological) conditions is important. In this case the content of volatile and gas components, which are lost at water extraction on the surface as well as the values of formation temperature and pressure, are of no particular interest. For this reason in the studies of consumer water properties the most attention was devoted to sample analysis methods and laboratory errors. In particular, all GOST's devoted to testing and quality control of waters for domestic/potable use render the most attention to sample reliability analysis and almost do not take into consideration a possible contamination at testing.

However, when the tasked issue has geochemical, exploration or ecological nature, its solution depends on water quality parameters directly in the geologic medium. In this case sought-for values must pertain to water properties and composition under the conditions of its natural occurrence. The completeness and reliability of these parameters directly depend on the preservation of the elemental composition of tested waters at their extraction onto the surface and on the accuracy of measuring their temperature and pressure in reservoir conditions. In this case the reliability of study results depends on the testing techniques, i.e., on all operations preceding measurement of water properties and composition in the sample.

Even at the most reliable repeated testing of the same water measurement, results of the same parameter will be different. This is a consequence both of natural variability of underground water properties and composition and changes in technical conditions of its testing. Natural variability is caused by natural processes and determines the minimum unavoidable error value, which does not depend on the testing techniques. The excess of these errors is associated with technical intervention in the process of underground water extraction on the surface, its testing and analysis. If we would like to describe variations of measured parameters by the value of variance σ_i^2, it may be presented as the sum:

$$\sigma_i^2 = \sigma_{i,\text{nat}}^2 + \sigma_{i,\text{tech}}^2, \tag{1.63}$$

where $\sigma_{i,\text{nat}}^2$ is variance of natural variation of the parameter i, $\sigma_{i,\text{tech}}^2$ – variance of technical variation of the same parameter associated with the extraction and study of ground water qualities.

With increasing volume of the geologic medium the number of natural factors affecting water properties and composition increases, and the size of variance $\sigma_{i,\text{nat}}^2$ grows. Natural variability of water properties and composition within the entire tested object may be the subject of study. For this reason it is very important for technical variations to be much lower than these natural macro-variations within the entire object $\left(\sigma_{i,\text{tech}}^2 << \sigma_{i,\text{nat}}^2\right)$.

On the other hand, as it is impossible to enter the same river twice so it is impossible at the same testing point to take one and the same water twice. Water sample properties and composition parameter variability at one point of testing determine the minimum variance value $\sigma^2_{i,\text{nat}}$. This value is usually disregarded. But these micro-variations also affect the value σ_i^2 and form the natural background (noise), which hinders the study of macro-variations. These natural micro-variations in the value of ground water properties and composition determine the natural component of measurement errors $\sigma^2_{i,\text{nat}}$, which is impossible to eliminate or lower. Even if $\sigma^2_{i,\text{tech}} = 0$ the value $\sigma_i^2 \neq 0$. According to equation I-63, with the decrease in testing's technical errors $\sigma^2_{i,\text{tech}}$ the role of the natural component $\sigma^2_{i,\text{nat}}$ grows. That is why it does not make sense to decrease technical testing errors $\sigma^2_{i,\text{tech}}$ below micro-variations $\sigma^2_{i,\text{nat}}$ at the testing points.

Maximum error acceptable at measurements of water quality and composition parameters is called ***limiting error*** or ***margin of error***. Any error, as a probability value, is characterized by some probability of its presence. Acceptable probability of margin of error is called ***confidence level***. Its value *P* describes the probability that sought-for genuine significance of a parameter does not exceed the set margin of error. The probability of error, i.e., that a sought-for number is outside the margin of error, is equal to the difference 1-*P* and is called ***significance level***. At the confidence level 0.95 the significance level is equal to 0.05. The higher the value of the confidence level, the lower the probability to make error. Usually standards of errors are assigned by the standard documents (GOST's, SNiP's, methodical instructions, etc.). Confidence level of these standards, as a regular rule, is set to be equal to 0.95.

In order to observe standards of errors at the set confidence level, control of testing quality is needed. Such control is implemented with the help of control samples collected at different stages of hydrogeochemical testing. With the growth of liability for the study results and decrease of acceptable margin of error, significance of such control and the number of necessary control samples increase.

2.2 Logistics of Field Testing

Fieldwork (field research) is the main, most responsible, most labor-intensive and, as a regular rule, most costly link in studying ground water composition. Often this research is done in a limited time, far from civilization and under uncomfortable conditions. Both its success and efficiency depend, first of all, on expertly worked logistics, which encompasses developing of the testing plan (testing time and points), providing access to underground water (wells, sampling pits, etc.) and analytical base and, at last, compiling the testing protocol. Besides this, the logistics also includes procurement of personnel, transport, equipment, instruments and chemicals.

The final objective of field work preparation is organizing the hydrochemical testing of underground waters at a specific geologic object with the assigned reliability, within the assigned period and at minimum financial costs. This preparation includes 5 major steps:

1. familiarizing with natural conditions and previous studies in the area;
2. planning of the regime and points of ground water testing;
3. preparation of wells and equipment for ground water extraction;
4. preparation of analytical base;
5. compiling of the testing protocol;
6. sample safekeeping and delivery to the laboratory.

2.2.1 Natural Conditions and Previous Studies of the Area

Natural conditions are defined by climate, topography, hydrography, and the geologic and hydrogeological environment.

Climate data enable the correct selection of the time or regime of hydrochemical testing. The amount of atmospheric precipitation, hydrography and topography provide first information about the nature, distribution and amount of natural water seepages suitable for testing (springs, swamps, rivers, etc.). Under conditions of humid climate ground water seepages on the surface are associated with lows and river valleys and are practically absent on water divides, in charge areas. Under the conditions of arid climate natural similar seepages are associated with endorheic (no-drainage) depressions and rarely with a discharge of artesian waters. In the population settlements are usually found artificial, equipped ground water seepages on the surface (shaft wells, karizs, drilling wells, etc.). Under the conditions of arid climate these seepages have especial significance and are often shown on physiographic maps. For artificial ground water seepages, data of their outfitting (depth, equipment and capitation design, status, etc.) are collected . At the shortage of water points ready for testing, especially in studies of hydrogeological objects at depth or for organizing the regime observations, additional special wells may be required.

The information about geologic and hydrogeological environment is necessary in order to correctly determine the object's complexity, to identify its relatively uniform areas and to estimate the needed testing density and uniformity. Hydrogeological stratification of a 3D object would enable a correct determination of the testing position in the section. Hydrodynamical data (hydrostatic heads, their gradients, water conductivity and filtration factors) enable determining the direction and velocity of ground water flow within the studied object.

At last, great value is associated with the results of previous hydrochemical studies, which give an idea of the ground water properties and composition. These data are especially useful for planning of testing points and substantiation of requirements to the reliability of analytical studies. With the availability of hydrogeochemical studies, which previously solved the assigned task, a ***map of previous hydrogeochemical studies*** is made.

2.2.2 Planning the Testing Regime and Points

Selection of the testing points and frequency is the most important and responsible step on the way to representative sample selection. The testing regime and points are planned in accordance with task assignment in consideration of physiographic, geologic and hydrogeological conditions but also of the availability of the existing water shows.

Regime testing/monitoring depends, first of all, on the regime of the studied object. At single-time testing a period is selected with minimum amount of the atmospheric precipitation (usually summer). Only under conditions of arid climate in the presence of secondary salinization such testing is sometimes more informative in a rainy period.

In monitoring studies, the frequency of testing the same water point has conceptual significance. In this case of conceptual significance is the monitoring objective. It is possible to identify three basic groups of monitoring tasks: 1. hydrochemical, 2. water quality control, 3. distribution contamination control.

In the first case is studied the hydrochemical regime of underground water; testing periodicity is determined by the nature of the studied natural factors of this regime. In connection with this in studying ground water regime the conceptual significance has: which fluctuations of water composition and properties (daily, seasonal or multi-annual) are planned to study.

For water quality control, monitoring is associated directly with the points of water collection. In this case the guides in the selection of testing frequency are, first of all, the law and the standards. In Russia the drinking water quality is controlled by the RF State sanitary-epidemiological service in accordance with its standard and methodological documents under (regularly the plan) and due to sanitary-epidemiological indications. According to SanPiN 2.1.4.1074-01 such control is provided by individual or legal owner under the set work program. At that, the number of ground water testings per year for organoleptic and micro-biologic

Table 2.2 Number of testings per year and their periodicity at potable water quality control depending on the population (SanPiN 2.1.4.1074-01).

Parameters	Population provided by this water-supply systemoй, thousands.		
	Less than 20	20-100	More than 100
Microbiological	50 (weakly)	150 (3 times a week)	365 (daily)
Organoleptic	50 (weakly)	150 (3 times a week)	365 (daily)
General parameters	4 (each season)	6 (once every 2 months)	12 (monthly)
Inorganic and organic matter	1	1	1
Radiological	1	1	1

parameters must be at least 4 (by the season) and for chemical and radiological parameters at least 1. For major water intake systems these requirements depend on the size of the population, which they provide with water (see Table 2.2). If the population is less than 20 thous. and in the absence of means of decontamination, underground water testing for organoleptic and micro-biological parameters is conducted at least once a month. For periods of high water and emergency situations, and also at the discovery of significant deterioration of water quality, a special testing/monitoring regime is established coordinating with the State Committee for Sanitary and Epidemiological Oversight center.

Control of contamination distribution is provided by observations of ground water property and composition changes in time at different testing points in the specific territory, sometimes also at different depths. This, as a regular rule, is the monitoring of several observation wells whose location and testing frequency depends on the nature of contamination focus and scale, on the pollutant nature and on hydrogeological environment. At control ground water monitoring around ecologically dangerous objects the frequency testing must be sufficient in order not to miss periodic contamination associated with emergencies or waste release. In the case of slow and long-term ground water contamination, the testing frequency depends mostly on the nature of the pollutant. M.J. Barcelona et al. attempted to estimate the testing frequency in consideration of Fisher information loss. Testing frequency derived by them with the loss of Fisher information under 10%, depending on the nature of components, is shown in Table 2.3. For many chemicals one testing every 3 months is sufficient for the description of short-time changes in time. For some components

Table 2.3 Calculated testing frequency (months) of water quality parameters with acceptable loss of Fisher information of no greater than 10% (Barcrelona et al., 1990; Site Characterization, 1991).

Parameters	In the absence of contamination	In the presence of contamination	
		increasing	**Declining**
Hydrochemical			
Macro components	1–2	7–14	1–5
Micro components (<1.0 mg/l)	1–2	<2	1–5
Parameters qualities waters			
Macro components	2–7	2–38	2–10
Micro components (<1.0 mg/l)	2–7	1–2	2–10
Parameters of water contamination			
Specific electric conductivity	6–7	24	7
pH	2	2	1
TOC	2	3	3
Content of organic halogen	6–7	24	7

actively reacting to oxidant-deoxidant environment (such as iron), the testing should be twice a month. For organic and radioactive components the testing frequency is set in accordance with their half-life.

Testing points depend on the type of studied hydrogeological object. In the case of a local object points of testing are determined directly by the assignment. In the case of regional and 3D hydrogeological objects, locations of testing points are planned before the beginning of field work. This planning depends on the assigned issues and the nature of hydrogeological object.

The plan of testing points location contains density and uniformity of testing point distribution. ***Sampling density*** is usually expressed in the value of average distance between testing points and ***sampling uniformity*** is usually expressed in the variance of these distances from their average values. The more complex the object and more responsible the task, the higher must be sampling density.

The simplest are *uniform objects* (***homogeneous entities***), for testing of which is required minimum number (1-3) of sampling points. Sometimes for the judgment of water properties and composition in such an object, a high-quality sampling of one water point is sufficient. It is important for the sample volume to be representative, i.e., to provide reliable information about the water properties and composition of the studied uniform object. Minimum volume sample, which reliably reflects the underground water properties and composition in geologic medium, is sometimes called ***minimal representative volume.***

But within most studied hydrogeological objects the ground water composition and properties noticeably, sometimes drastically change. These changes are caused by geologic, hydrogeological and technogenic environments and often are the object of study. Such ***nonuniform (inhomogeneous) objects*** require zonal testing. First of all they are stratified, i.e., separated into portions that are relatively uniform in the environment and ground water composition (different water horizons, their blocks separated by faults, etc.). The greater the number of such uniform but different parts that may be identified within a hydrogeological object, the more complex it is as a whole. The number of water points necessary for achieving the required reliability in nonuniform objects increases, as a regular rule, in proportion to their complexity.

Several different techniques exist for planning ground water sampling points. Disorderly, zoned, linear, orderly, disorderly by grid and exploratory planning is identified.

Random location (Figure 2.1a) is usually used for testing ground water over large territories limited by the existing water-shows (springs, seepages, swamps, shaft wells, drilling wells, etc.). As a regular rule, it is not planned. It is used in hydrogeological surveys, in hydrochemical exploration for economic deposits or in preliminary reconnaissance of the hydrochemical environment. Such planning of the testing is acceptable only for small-scale regional studies and for relatively uniform objects, within which underground water properties and composition practically do not change. Density testing in these cases depends on the mapping scale and complexity of the object of study. Density testing recommended by A.A. Brodsky (1957) for exploration hydrochemistry, depending on the mapping scale (Table 2.4) and geologic complexity may be an example.

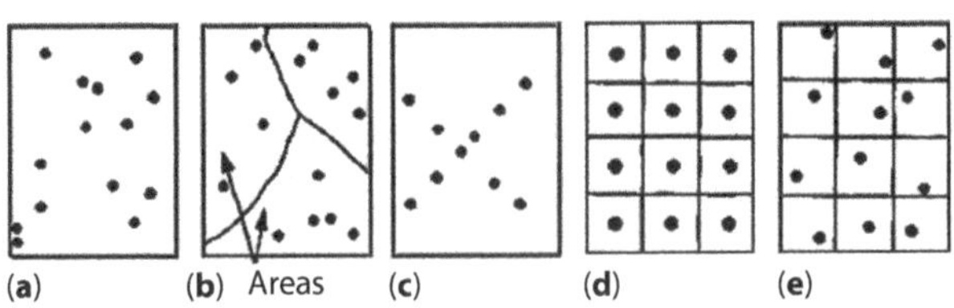

Figure 2.1 Some location patterns of testin points in regional studies: a. simple random, b. zoned random, c. linear, d. orderly by grid, e. disorderly by grid.

Table 2.4 Number of points of regional testing per 1 km^2 depending on scale of mapping and complexity of the geologic environment* (A.A. Brodsky, 1957).

Mapping scale	Geologic structure		
	Simple	Medium complexity	Complex
1: 1,000,000	0.015	0.02	0.04
1:500,000	0.03	0.05	0.08
1:200,000	0.09	0.13	0.2
1:100,000	0.3	0.45	0.6
1: 50,000	0.7	1.0	1.6
1:25,000	1.8	2.6	3.6

*Complexity parameters are determined under the guidance of the *Reference book of aggregate budget standards exploration operations*, issue 1. Gosgeoltekhizdat, 1954.

In more responsible and complex cases (engineering or ecological works), when it is impossible to use only testing of ground water shows, the points for shaft wells or drilling wells are planned. Their number and locations are determined at the stage of work preparation. Planning of their locations begins with the stratification of the study object, i.e., with dividing it into areas, within which changes in hydrogeological conditions may be disregarded (Figure 2.1b). Within each uniform portion of the hydrogeological object, location of the testing points depends on its size. In small portions it is possible to have just 2-3 points. In large territories two principles of orderly planning are applied: linear and grid. Under the linear one, testing points are positioned along the lines running along or across the ground water flow direction (Figure 2.1c). Under the grid planning, the territory is parcelled into a system of equal squares, and testing points are distributed uniformly between the squares. The sampling points within squares are spotted either arbitrarily (Figure 1.18d) or exactly (in the center or at the junction of the corners) (Figure 2.1e). Such methods are used for detailed studies of ground water property and composition changes when it is necessary to estimate boundaries of the contamination halo and a direction of its migration. When testing points are planned in advance, a ***map of planned testing*** is made.

Sometimes exploratory testing is used. It is customary in the search of economic deposits or contamination foci and begins with the selection of 3 points of testing. Based on the results of ground water properties and composition study, at these points are set up additional testing points placed in the direction of the concentration indicator component increase.

The type of sampling point locations does not affect the reliability of testing results for individual points but may render noticeable influence on the reliability of hydrochemical description of the geologic object as a whole, on average property and composition parameter values. Unfortunately, this influence caused by an unhappy location of testing points is difficult to estimate. Obviously, the greater the density and uniformity of sampling, the more complete the use of information about physiographic, geologic and hydrogeological environment, and the lower the error of a hydrogeochemical description caused by the sampling point distribution.

2.2.3 Preparation of Wells and Equipment

Not all ground waters are exposed on the surface. In most cases, in order to penetrate and sample them it is necessary to dig shaft wells or to drill wells. The deeper the water occurs, the more technically difficult it is to extract it on the surface. The greatest difficulty is associated with testing of deep ground water in wells when at extraction on the surface the environment drastically changes. Besides, on its way to the surface the underground water may interact with materials of the wells, equipment for its extraction, and with air, oxygen and other foreign matter. All factors, affecting the composition and properties of the water in its testing, may be subdivided into 4 groups: 1. Change of environment, 2. Interaction with materials of the wells and lifting equipment, 3. Well design deficiencies, 4. Particulars of water lifting techniques.

Influence of the change of environment on the ground water composition needs to be taken into account if collecting samples from depths greater than 10 m. At extraction from these depths and greater, the water may imperceptibly change its properties and composition due to: a. changes in temperature and pressure, b. interaction with air oxygen, c. catching of foreign water.

a. The decline in temperature and pressure facilitates decrease in solubility of some components, and noticeably affects properties and composition of waters close to the saturation. Declined pressure may cause water degassing, which is sometimes manifested by the release of spontaneous natural gas. Most commonly released is CO_2. This causes pH decrease and imperceptible carbonate precipitation. Lowered temperature may cause salt precipitation from brines. Pressure and temperature influence is often possible to notice visually and to eliminate by improving of the testing technique, in particular using special hermetic samplers.

b. The interaction with air oxygen may affect the content of components sensitive to change in oxidant-deoxidant environment (polyvalent metals, organic matter, NO_3^-, NO_2^-, NH_4^+, H_2S, etc.). However, oxidizing processes are quite slow, and for this reason their influence is possible to limit by a decrease in the area and time of water contact with air.
c. Catching water and matter of foreign sources is associated either with poor well flushing after its equipping or with its insufficient eë hermeticity. This may be drilling fluid, stagnant water or water from an outside horizon. To eliminate the effect of materials remaining after drilling and equipping, the wells are carefully flushed until properties and composition of underground waters stabilize. The introduction of the outside ground water is associated either with too large a filter or with insufficient hermeticity of the borehole. This is possible to eliminate only by careful treatment of well design and status.

A long filter mixes ground waters from a wide depth range (Figure 2. 2), and these waters may be substantially different in properties and composition. Quite thick sand horizon in Figure 2.3 may be an example. Its lower section is saturated with water with negative Eh and with Fe^{2+} content. Its upper section is saturated with water with positive Eh and with O_2. In testing of this horizon with one filter the waters mix, and the following reaction occurs:

$$2Fe^{2+} + 0.5O_2 + 3H_2O \rightarrow 2FeOOH + 4H^+.$$

As a result the sample may contain simultaneously Fe^{2+} and O_2, but its composition, Eh and pH, will not reliably reflect ground water properties and composition in the tested horizon. The same error is especially

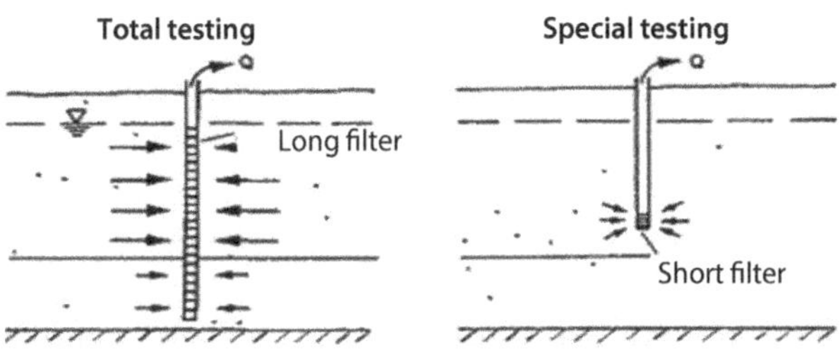

Figure 2.2 Water sample representativity vs. well design in conditions of different rock permeability. Size of the arrows reflects flow velocity.

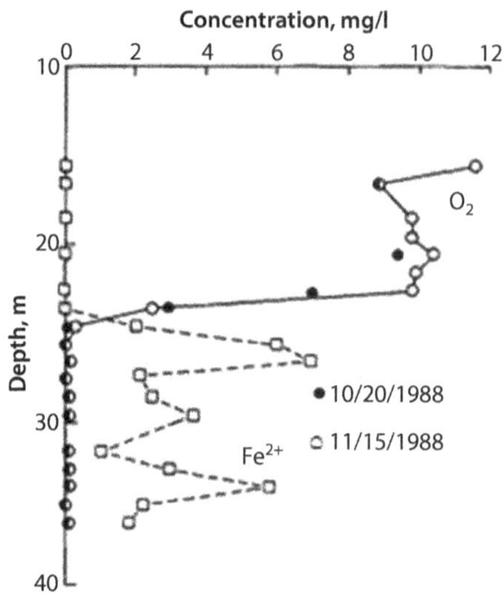

Figure 2.3 O_2 and Fe^{2+} distribution in water-saturated sand horizon.

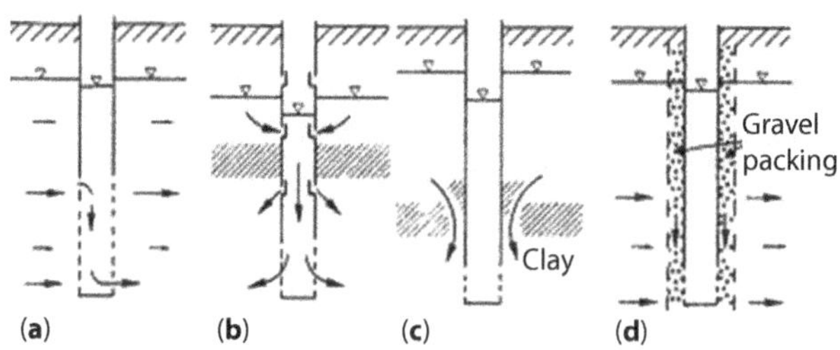

Figure 2.4 Different forms of water cross-flow at testing due to imperfections in well design. A. Very long filter, B. damaged casing joints, C. Incorrectly placed bentonite packer accepts cross-flow behind the casing, D. Gravel packing facilitates connection between horizons.

dangerous when penetrating several water horizons. Examples of crossflow and mixing of ground water from different horizons, associated with technical deficiency of wells, are shown in Figure 2.4. The cunning of errors with catching waters of different composition is that they become noticeable only after analysis of samples, usually at comparing of the obtained results. Elimination of these errors is possible only through a very strict control of the design and state of the tested wells. For this reason, at testing of several water horizons, it is not recommended to use very long filters and old wells. Sometimes it is beneficial to observe changes in water composition under different regimes of well drawdown.

Interaction of the water with well materials and lifting equipment may influence the content of mineral and organic micro-components. First, wells and water lifting equipment may be dirtied by substances used in well drilling and equipping. Moreover, the multiple devices used in testing may contain the traces of previous samples and thereby may cause cross-contamination. In order to eliminate this phenomenon the wells are flushed, meaning the water lifting equipment is carefully and numerously washed. Clean materials may also interact with the tested water and affect its composition. These materials are capable of exchanging components with water due to corrosion, leaching, solution and sorption. The influence of such mass-exchange increases when water has high temperature, low pH values and corrosive components. Therefore, in special testing for micro-components or organic compounds it is necessary to be very careful to the materials used for wells and lifting equipment.

On the other hand, all materials of the casing, tubing and filter must be quite durable. In many cases they must provide for well operation without corrosion for at least 30 years. At the same time they must be inert to underground water and not affect its composition. Compromise between these requirements usually provides such solid materials as stainless steel and some polymers: polytetrafluorethilene (PTFE or Teflon) and polyvinylchloride (PVC). No less important are the materials used directly in testing (crates, tubs, etc.). In testing for micro-concentrations of metals it is recommended to use stainless steel or strong and chemically resistant polymers. For organic components, stainless steel, aluminum and polystyrene should be selected. Materials used in ground water testing and their advantages and drawbacks are listed in Table 2.5.

In real life for ground water testing it is often necessary to use wells not intended for this purpose. Most of these wells are associated with exploration and appraisal of economic deposits and production of underground water. Sometimes samples are collected from unequipped wells in the process of their drilling. The sampled wells may be subdivided into 4 main groups:

1. Uncased; samples are selected right in the borehole,
2. Exploratory and appraisal,
3. Production,
4. Special.

The simplest and cheapest way of ground water testing is directly in the borehole at water penetration while drilling. In this case, however, depths of sampled intervals are often determined unreliably, the necessary quality of

Table 2.5 Materials for well equipment, sample collection instruments and vessels.

Material	Advantages and drawbacks
Cast iron and low-carbon steel	Very strong, fireproof, chemically passive but brittle, heavy (density 7.0 –7.8 $g \cdot cm^{-3}$), easily rusts and destroyed under corrosion environment. Serves a source of outside metals (Fe, Mn, Cd, Cr, Cu, Zn) and good adsorbent for metals in water
Galvanized steel	Very strong, fireproof but rigid, heavy (7.7 – 7.9 $g \cdot cm^{-3}$), may be source outside metals (Fe, Cd, Mn, Zn), especially if the coating is damaged and rust appeared, and adsorbent for many metals in waters.
Stainless steel	Very strong, fireproof, chemically resistant but heavy (7.9 – 8.2 $g \cdot cm^{-3}$), at low pH a source of outside metals (Fe, Mn, Cd, Cr, Cu, Zn), capable of being a catalyst in organic reactions.
Brass	Plastic alloy of high density (8.3 – 8.7 $g \cdot cm^{-3}$) and strength, resistant to corrosion, under normal conditions is not oxidized in the air but easily reacts with halogens, sulfur, may be a source of Si, outside metals (Cu, Zn, Sn, Ni, Pb, Mn, Fe), etc.
Glass	Quite strong, fireproof, 2.4–2.8 $g \cdot cm^{-3}$, chemically very passive but brittle, introduces SiO_2, B, Ca, Mg, Na, K, Fe, etc.; adsorbent for metals.
Polytetrafluorethylene (PTFE) or Teflon®	Light fluoroplast (2.2 $g \cdot cm^{-3}$) fireproof between –70 and +205°C, elastic with high impact viscosity, resistant *to* acids and alkali, oxidants, not wettable by water or most organic solvents, but the strength is lower than of steel and cast iron, source of F.
Fluorinated ethylene-propylene (FEP)	Light fluoroplast (2.15 $g \cdot cm^{-3}$), fireproofостью –40 to +204 °C, elastic with high impact viscosity, resistant to acids and alkali, oxidants, nonwettable by water or most organic solvents but the strength lower than of steel and cast iron, source F.
Polyvinylidenfluoride (PVDF) or Kaynar®	Light fluoroplast (1.77 $g \cdot cm^{-3}$), fireproof to +150°C, elastic with high impact viscosity, resistant to acids and alkali, oxidants, nonwettable by water or most organic solvents but strength lower than of steel and cast iron, source F. Poor chemical resistance to ketones, acetone.
Polyvinyl chloride (PVC)	Light chloroplast (1.35-1.43 $g \cdot cm^{-3}$), elastic, quite low permeability to liquids, vapors and gases, moderately resistant to alkali, many acids, mineral oils (lubricants) and solvents but has limited fireproof capacity (-15 to 65 °C), destroyed on contact with ketones, esters and aromatic hydrocarbons, source of Cl.

Polystyrene (PS)	Light copolymer (1.05 g·cm^{-3}) fireproof –40 to 105°C, low humidity absorption, good resistivity to even concentrated acids (except nitric and anhydric acetic) and alkali, resistant to radioactive radiation but brittle, soluble in own monomers, aromatic hydrocarbons, esters, acetone, ages in the air under UV radiation, develops micro-fractures, increased brittleness, may introduce organic carbon.
Acrylonitrile-butadienestyrene copolymer (ABS)	Light copolymer (1.06 g·cm^{-3}), fireproof –40 to 105°C, rigid, brittle, low mechanical strength, resistant to diluted alkali, acids, volatile organic but not resistant to acetone, ether, benzene, ethylchloride, ethylenchloride. May introduce organic carbon and As, adsorbs metals (Zn, Li, Hg, Mo, Se).
Polyethylene (PE)	Light polymer (0.916–0.960 g·cm^{-3}) fireproof –60 to 80°C, with low steam- and gas-permeability. Resistant to water, alcohol, gasoline, does not react with alkali of any concentrations, with solutions of any salts, carboxylic, concentrated hydrochloric and fluoric acids but has low strength, sensitive to UV radiation and heat, decomposes on contact with 50% nitric acid at room temperature and with chlorine and fluorine. May introduce organic carbon and adsorbs metals.
Polypropylene (PP)	Light (0.900-0.910 g·cm^{-3}), fireproof –10 to 140°C, resistant to mineral acids, alkali, salts solutions, moderately resistant to alcohols, ketones and esters but highly sensitive to light and oxygen, low strength, destroyed on contacte with oxidizing acids, aliphatic and aromatic hydrocarbons, soluble in some solvents, for instance, in benzene, tetrachlormethane, ether. May introduce organic carbon; adsorbs metals.

samples is not provided for, and there is no possibility of their repeated collection. At penetration of several water horizons it is possible to use *packers*, special devices for their separation in the borehole. When the water shows up, the drilling is stopped, and a packer is run into the well, isolating the upper part of a well form the bottomhole, which allows for testing and sampling only the penetrated horizon. After the test the packer is extracted and drilling continues. This technique does not accept preliminary well flushing and does not guarantee purity of the sample. Analyses results of such samples may be used only as an addition to more reliable data.

Exploratory and appraisal wells are sometimes accessible for hydrochemical testing but only for a short period of their testing. Conditions of their testing depend on their purpose and facilities. In the presence of casing in shallow wells, as a regular rule, there is a filter through which the water is extracted and sampled. It is important to know the depths of occurrence and age of the penetrated horizon (rock), well casing design, and casing and filter material. The deepest ground waters are penetrated in most cases in wells intended for exploring fields of oil and gas. They are tested after drilling by means of the perforation. For this the casing is run to the bottomhole and the annulus is cemented. For the study and testing of the lowermost reservoir, the casing is perforated at the depth of occurrence. It is done by a perforator, which punches through the casing and cement behind it and creates additional fractures in the reservoir. After the testing the perforated interval is isolated by setting a so-called cement bridge and the perforation of the next interval up the borehole is conducted. The inflow of water sometimes is caused by lowering pressure in the well or the other way around, by its increase, i.e., hydraulic fracturing of the reservoir (fracking).

The exception is exploitation wells for water-supply, which may be accessible for testing for several years. However, the design features of these wells do not always allow for sample collection at their wellhead in accordance with requirements of qualitative hydrochemical testing. For this reason before using such wells it is necessary to become familiar with their geologic column, casing design, the status at the moment of testing, the water-supply system and, of course, with their owner.

If old wells are intended to be used for testing, it is necessary to become familiar in advance with their state. The wells may be suspended, temporarily abandoned or even plugged and abandoned. Not all of them penetrate and exploit water horizons needed for testing. In Russia low-grade steel and cast iron were often used for well facilities; they are subjected to corrosion and introduce Fe, Mn, Cd, Cr, Cu and Zn in the water. Sometimes copper or brass filters are used, from which may come Cu and Zn. In most cases for access to testing of the existing wells, the owner's consent is required, and the owner may be located far from the well.

Special wells are drilled and equipped directly for hydrogeochemical studies. Their main design parameters are depth, performance capacity, size and materials of their casing and filter. For regional studies temporary wells are drilled; after testing they are often plugged and abandoned. Drilling and equipping these wells is done before the beginning of field operations. Special ***monitoring*** (***observation***) ***wells*** are equipped for regime studies. They are equipped for a long period in consideration of the objectives and tasks of hydrochemical testing. For this reason special attention in their equipment is paid to isolation of tested horizons and to the quality of used materials. The wells must be durable and free of components studied in ground waters.

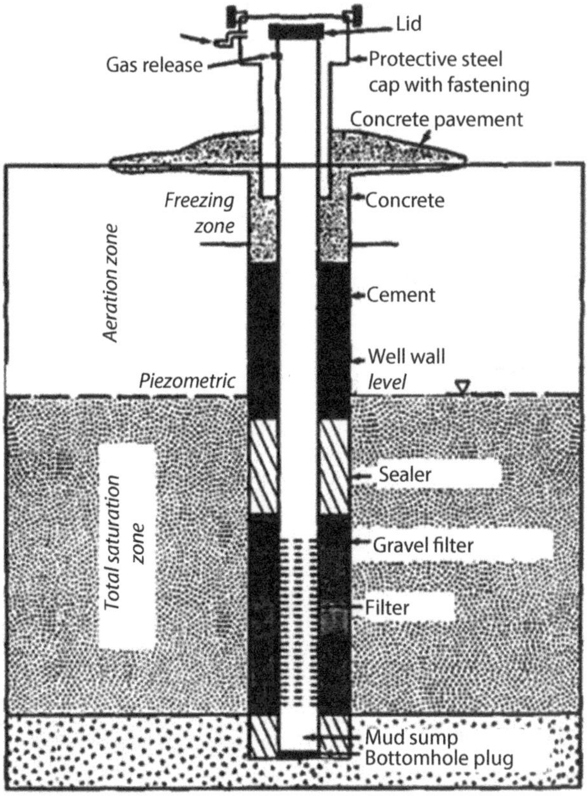

Figure 2.5 Typical monitoring well with casing and filter from stainless steel.

As a regular rule, observation wells are completed for one water horizon. Figure 2.5 shows an ideal observation well of stainless steel for repetitive hydrogeochemical studies of an individual horizon. In a case of a 3D hydrogeological object it is necessary to conduct observations of changes in the water properties and composition with depth. For this purpose either individual wells or simultaneously several wells are used. Figure 2.6 displays main methods of setting up a monitoring ground water testing at different depths at one point: 1. With a packer, 2. With connected wells - piezometers, 3. With a (piezometer) well cluster of different depths.

Packer is a special device for isolating one part of a well from the other. Mostly used are Inflatable packers. They are run to needed depths, inflated and isolate the lower part of a well from the upper one. Two packers enable the separation and sampling of the well section part by part, for which purpose wells are equipped with a long filter. Obviously, such testing is

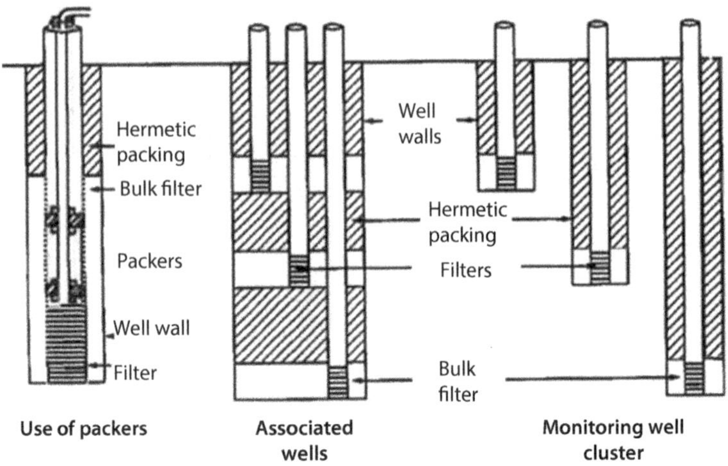

Figure 2.6 Three techniques of setting up hydrochemical testing with depth.

no guarantee against crossflows along the filter. For this reason the use of packers is possible only at relatively low rates of flow.

More reliable isolation of water at different depths in one productive horizon is achieved by using several wells. For this purpose conjugated small diameter piezometers placed in a large diameter well are sometimes used. In this case sampled intervals are isolated from one another by concrete or cement packers. With good isolation this technique may be used also for regime observations. Nevertheless, for a reliable study of ground water property and composition changes with depth it is recommended to use clusters of observation wells, which penetrate different depths of one horizon or even different water horizons.

Equipment for water lifting. In testing of deep ground water, especially when working with wells, it is necessary to prepare in advance tools and equipment for extracting water. In selecting the tools one should be guided not only by their size, simplicity of operation, availability and cost but also by the task assignment, depth of the water occurrence, and its composition and design features of the tested wells. There are two conceptually different ways of extracting the underground water on the surface in the absence of a natural flow: with pumps and samplers.

Pumps for wells may be downhole (lowered in wells) and surface (installed on the surface), centrifugal, vibration and vortex. They all require hoses or tubes, source of electric power, etc. Their selection is determined not only by technical parameters, weight and price but also by the task assignment, depths of underground water occurrence and well design. Pumps provide for very comfortable conditions and unlimited volume of water for collection. But their use is relatively expensive and justified only for long testing and sufficient depths of ground water occurrence, in particular for flushing of recently completed or abandoned wells. Most water-intake wells are equipped with pumps. Water extraction from great depths with a pump is no guarantee of preservation of its elemental composition, as during lifting, due to the drop of pressure, most volatile organic and gas components may be lost. At high salinity some mineral components may be lost also. For testing volatile components and the determination of the gas-water ratio values at a natural flow or forced withdrawal, it is possible to use a gas separator (Figure 2.7).

Samplers provide for a simpler method of limited volume of water. In case of a shaft well a simple bucket may be the sampler. For testing a well, devices of small diameter are necessary. In shallow wells the water is lifted with a bailing tube (Figure 2.8, a) or other simple devices (Figure 2.8, b, c, and d). All these devices, however, are applicable for testing waters at depths up to 10-20 m. Testing with these devices at great depths carries the risk of losing volatile organic and gas components. For this reason for testing deep ground water, special downhole samplers should be used. They are run on a steel cable, open to the required depths where they are hermetically closed and recovered together on the surface with

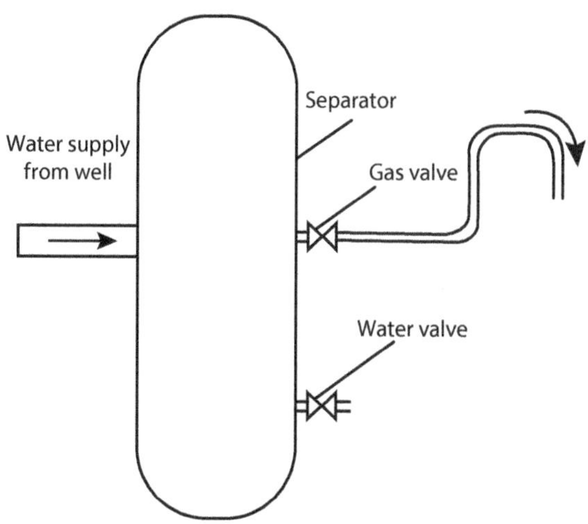

Figure 2.7 Gas separator, in which the water accumulates at the bottom and the natural gas, at the top. A water sample is collected through the lover valve, and a natural gas sample through the upper valve. The ratio of gas and water rates under the stabilized withdrawal regime determines the value of gas-water ratio.

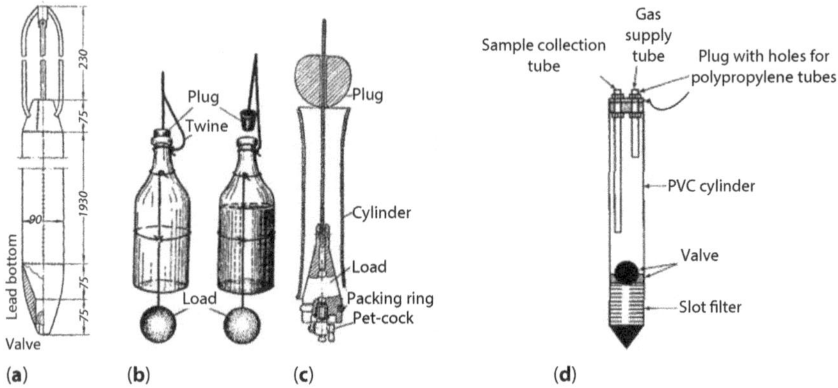

Figure 2.8 Devices for underground water extraction from wells: (a) Bailing tube, diameter 90 mm and capacity 8 liters. When the bailing tube is run down the hole, the valve of the conical bottom is open and water enters it. When the bailing tube is run up, the valve is closed and is retaining the water to surface. (b) Bottle, empty and plugged, is run to a needed depth where the plug is jerked out. The filled-up bottle is raised on the surface. (c) E.V. Simonov's device. First the load with the packing rig and petcock is run on a steel wire cable to needed depth, then the cylinder and then the heavy plug. (d) Testing by gas pumping. At needed depth an inert gas is pumped. Due to this the valve at the sampler's base closes and the water is pushed out to the second tube and up.

sample water. Some time ago in Russia, E.V. Simonov's telescopic sampler was commonly used. Unfortunately, it did not have sufficient hermeticity and lost gas components. There are more reliable samplers, which maintain hermeticity under normal conditions even at a great internal pressure. In particular, samplers PG-1000 and VPP-300 enable collecting 0.4-0.9 liter samples of water and crude oil practically without loss of the dissolved gas from the depths of up to 3 km at temperatures up to 200°C and greater. Schematic design of one such sampler is displayed in Figure 2.9.

All equipment for ground water extraction is prepared and tested before the beginning of field operations. For water pumping from wells it is important to correctly select a pump, tubes and hoses of sufficient length and to provide electric power supply. To operate hermetic samplers at great depths it is necessary to have the truck with a pulling hoist and steel cable of sufficient length.

2.2.4 Preparation of Analytical Base

The major task of preparing analytical base is providing the study ground water properties and composition with necessary and sufficient methods

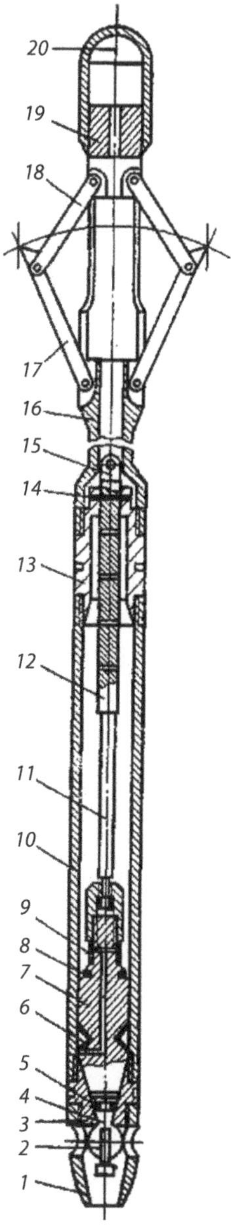

Figure 2.9 Schematics of a special plunger sampler GPV-4S.

The sampler is run on a steel cable 20, which through sliding block 19, pulling rod 11 and its lower head 12 is connected with plunger of valve 7. The sampler body 16 is attached to plunger's pulling rod with copper locking pin 14. The sampler is run down the hole and not flushed by water to the testing point. At the assigned depths it is slightly raised, levers 17 stick to the well walls and with the pull 18 tear away the copper locking pin 14. The sampler is released of the plunger pull and its body 16 goes down relative plunger 7. The plunger with the help of dual cup cuff 6 displaces water through the upper orifice and fills up the chamber 10 with the sampled water through the tip 1. Eventually the rubber ring 8 and valve 9 on the plunger enter the upper saddle 13, valve 4 and rubber ring 5 are pushed by the spring 2 to the lower saddle 3. While the sampler is being raised the excess of the internal pressure over external increases and locks the upper and lower valves thereby providing for hermeticity of the sampler chamber. (V.N. Kortsenshtein, 1991)

and means. At that, the guide is, first of all, the task assignment and already existing analytical material. The task assignment determines the nature and number of the composition parameters, which are necessary to analyze. The analytical material collected from publications and files allows for the preparation of the analytical base. First, it gives the idea of ground water composition and properties, which will be studied, thereby helping to determine the requirements to the analytical base quality. Second, this material provides an opportunity to estimate the reliability of the new data. And at last, third, it may be used as addition to the new results of the field study. In particular, this material may contain analytical data for water points inaccessible at the time of testing (dried up, abandoned, closed, etc.). Old analytical data are especially important for monitoring studies as they broaden the temporal range of observations. Their use, as a regular rule, is inacceptable for one-time local studies. However, the old data on the water composition and properties should be treated with great caution, especially in the absence of information about methods of testing, analysis and analysis performers.

On the whole, the analytical base preparation includes:

1. Determination of the measured property and composition parameters;
2. Determination of acceptable measurements errors;
3. Selection of techniques and means of chemical analyses;
4. Selection of points and performers of these measurements;
5. Providing the field study with necessary tools, chemicals, extractants, etc.

2.2.4.1 Selection of Property and Composition Parameters

Selection of property and composition parameters, which are necessary to measure, is determined by the task assignment. Nevertheless, practically in any testing organoleptic properties of waters and pH are determined and the general analysis is conducted in the field or laboratory. Also it should be recommended to determine electric conductivity value, which would give an idea of water salinity and may be used for subsequent control of analyses correctness. This minimum selection of parameters is actually included in any testing. Gas components are usually determined qualitatively (by flammability and smell) if they show up themselves. Other parameters are measured in special studies. In a case of searching for ore deposits, the main attention is devoted to heavy metals, which are search attributes. In the search of oil or gas fields, the attention is devoted to halogens, gas and organic components (See Table 2.6). In a case of exploring for domestic/potable waters, the first priorities are organoleptic properties and also parameters included under SanPiN 2.1.4.1074-01. Conceptual significance for ground water quality control has the nature of possible contamination foci. In industrial areas special attention is devoted to heavy metals and organic components in the industrial waste (petroleum products, surfactants, halogen-containing organics, etc.). In rural locations greater attention is devoted to nitrogen, phosphorus compounds, pesticides, etc.

2.2.4.2 Substantiation of Margin of Error Measurements

The most responsible operation in the hydrogeochemical testing is ***measurements***, which in hydrochemical studies is a complex process including both field testing and laboratory analysis. Most measurements are performed under variable conditions by several executants. Such process can be described as a relay race of experts of different disciplines who carry one baton, the sample, to a single objective. This objective is numerical values of ground water properties and composition within the restraints of the ***required accuracy***.

Measurement of the environmental parameters is attributed within the purview of the state regulations, which is provided by the RF Federal Law "About ensuring unity of measurements." Under this law measurement of ground water properties and composition must be conducted in accordance with certified techniques (methods) with their results expressed in units of value accepted for the application in the Russian Federation. GOST P ISO 5725-2002 "Accuracy (correctness and precision) of measurement techniques and results" was enacted 1 November, 2002. This standard corresponds with the international standard ISO 5725 regulating clauses and determination of concepts in estimating accuracy techniques and results of measurements accepted in the world.

2.2.4.2.1 Parameters of Testing Errors

According to GOST P ISO 5725-1-2002 all measurement of individual water quality parameters within the boundaries of a studied object must be performed by one standard technique within the framework of the assigned accuracy parameters.

Any measurements contain some technical errors, which are possible to estimate and lower by repeated measurements. Most frequently encountered value of measurement results, i.e., mode is considered the closest to the sought-for true value. The ***accuracy*** describes the extent of this

Table 2.6 Selection of ground water properties and composition parameters depending on the task assignment of hydrogeochemical testing.

Analyzed parameter	Points of analysis	Testing objective						
	Lab= laboratory, Field = field	Hydrogeological survey	Search for ore deposits	Oil and gas exploration	Quality of drinking water	Water quality control	Engineering-technical evaluation	Discharge of waste water in rivers
Temperature	Field	×		×	×		×	×
Organoleptic properties	Field	×	×		×		×	
Electric conductivity	Field	×	×		×	×		×
Radioactivity	Field	×	×		×	×	×	
pH	Field	×	×	×	×	×	×	×
Alkalinity	Field/Lab					×	×	
Eh	Field			×		×	×	×
Na+, K+, Ca^{2+}, Cl-, SO_4^{2-}, Mg^{2+}	Field/Lab	×	×	×	×	×	×	×
NO_3^-, NO_2-, NH_4^+	Field/Lab			×	×	×		×

Halogens	Lab			F, I, Br	F^-	F^-		F^-
PO_4^{2-}	Field	×			×	×		×
H_4SiO_4	Lab						×	
Salinity	Lab			×	×	×		×
Hardness	Field/Lab				×		×	
Heavy metals	Lab		×		Al, Be, Ba, Fe, As, Cd, Cr,Mn, Cu, Mo, Ni, Hg, Pb, Se, Sr, Zn		Al, Fe, Mn	As, Be, Cd, Co, Cr, Cu, Fe, Hg, Mn, Mo, Pb, Se, Te, V, W, Zn
Gas-saturation	Field/Lab			×			×	
Gas components	Field/Lab	Горючесть, H_2S		CH_4, CO_2, N_2, H_2S, Ar, He	O_2	O_2, CO_2, H_2S	CO_2	O_2
BOD, COD, TOC	Lab			×	×	×		
Organic components	Lab			×	×	×		
Content of halogen organics	Lab				×	×		×
Micro-organisms	Lab		×	×	×	×		×

Source: Modified from "Site Characterization for Subsurface Remediation." Seminar Publication

closeness. The measure of measurement accuracy is the ***absolute global error***, which is equal to the difference between the obtained modal and true values. In the case of concentrations the value of the global errors is determined from the following equation:

$$\Delta C_i = | C_i - C_{i,true} |, \tag{1.64}$$

where C_i and $C_{i,true}$ are measured and sought-for (target) true value concentrations, respectively. The higher the error, the lower the accuracy. Global error includes both precision and correctness.

Precision describes the extent of closeness to each other of independent measurement results of one parameter under the same conditions (Figure 2.10). This parameter depends only on chance factors and is not associated in any way with the true value. The measure of the precision estimate is ***mean square*** or ***standard deviation*** σ_i, which describes the scatter of measurement results relative to their average values:

$$\sigma_i = \sqrt{\frac{\sum_{n=1}^{N} \left(C_{i,n} - \overline{C}_i\right)^2}{N}}, \tag{1.65}$$

where $C_{i,n}$ is the result of nth determination; $\overline{C}_i$ is the mean value (mathematical expectation) of observed values C_i, N is number of measurements. Value σ_i^2 is called ***variance***. For a reliable determination of its value, a sufficiently large number of repeated measurements – N – is required. The variance is additive, and total variance of the measurement technique is possible to represent, as vector, by the summary variance of its individual factors:

$$\sigma_i^2 = \sum_{n=1} \sigma_{i,n}^2 \tag{1.66}$$

The greater the standard deviation or variance, the lower the precision. Their values are found depending on measurement conditions. Precision serves the measure of ***random error,*** which is equiprobable excess and underestimation of the measurements relative to their mode. Two basic precision levels are distinguished: repeatability (convergence) and reproducibility.

Repeatability or ***convergence*** – precision under conditions of observance the complete identicity of measurement techniques of the same parameter. ***Reproducibility*** describes precision at the use of different measurement techniques (different methods of testing, methods of analysis, etc.). Thus, repeatability and reproducibility describe the range of possible values of random errors, respectively, lower and upper. Random error is a consequence of subtle random variations, both natural and technical. For this reason random error cannot be completely eliminated. It may be only decreased.

Correctness (*trueness*) determines the extent of closeness of the measurement results to the sought-for true value (Figure 2.10). The difference between the measurement results and some accepted reference (ideal) value serves as a measure of correctness of separate measurements. ***Ideal values*** replace the true ones and may be: 1. theoretical or scientifically-established; 2. certified; 3. mathematically expected for the selection (average value of the assigned population of analysis results). At the presence of repetitive measurings the correctness is equal to the difference between the selection mode and ideal value. Correctness determines deterministic ***systematic error***, which describes either excess or underestimation of measured values relative to ideal values. Its value may be a function of measurement conditions or time, due to, for instance, change of executant, contamination of samples and chemical, etc. If this function is identified, the systematic error may be excluded.

Eventually global error of repeated measurements of the same parameter is the sum of systematic θ_i and random $\underline{\Delta}_i$ errors:

$$\Delta C_i = \theta_i \pm \underline{\Delta}_i, \tag{1.67}$$

and each measurement result, C_i, includes in itself three addends:

$$C_i = C_{i,true} + \theta_i \pm \underline{\Delta}_i. \tag{1.68}$$

Value of the absolute global error depends on the values of measured parameter, in particular, values of the average concentration, which may

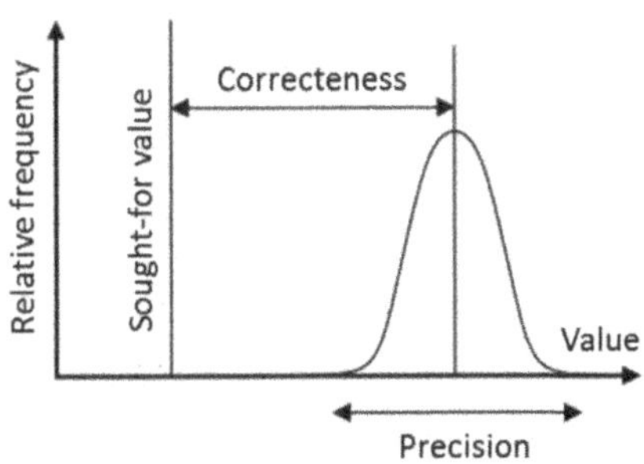

Figure 2.10 Interrelations of precision and correctness in estimation of errors measurements.

change within a rather broad range. Value of their ratio is called ***relative error***:

$$\delta_i = \frac{\Delta C_i}{\overline{C_i}} = \frac{\theta_i}{\overline{C_i}} \pm \frac{\Delta_i}{\overline{C_i}} = \underline{\delta}_i \pm \underline{\underline{\delta}}_i, \tag{1.69}$$

where δ_i, $\underline{\delta}_i$ and $\underline{\underline{\delta}}_i$ are relative errors: global, systematic and random, respectively, and $\overline{C_i}$ is average value of parameter *i*.

As $\overline{C_i}$ average values approach zero, measurements of relative global error drastically increase and reach some limit, below which measured parameter values lose their informative meaning. At concentrations so low, it is possible to discover components where there are practically none, and on the contrary, not to notice them where they are. As many components are present in ground waters in micro-concentrations, it is very important to have an idea about the minimum concentrations, which testing technique is capable of discovering or quantitatively determining. Minimum analyte concentrations, which the technique allows to discover or quantitatively measure, describe the technique's ***sensitivity*** as a whole. Two sensitivity thresholds are distinguished: limit of detection (DL) and limit of quantification (QL) (Figure 2.11).

Limit of detection of the analyte describes the lowermost concentration, which may be determined as statistically distinguishable from the concentrations of *blank samples*, which would prove the presence of the analyzed component with assigned statistical reliability. *Blank samples* are water samples that went through the entire measurement process but originally did not contain the needed analyte. Usually this is distilled and nonionized water. In Russian Federation for this purpose is used, as a regular rule, distilled water with electric resistivity of at least 0.002 megohm·cm^{-1} (GOST P 6709-72 Distilled water. Technical conditions), in the USA – at least 1 megohm·cm^{-1}. For organic components sometimes the water purified with high-efficiency liquid chromatography is used. Blank samples are commonly used at different stages of testing for the

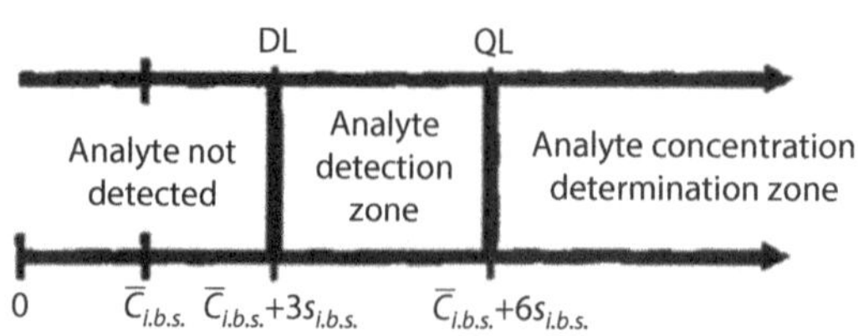

Figure 2.11 Minimum analyte concentrations, which used testing methods allow detecting or measuring quantitatively.

determination of not only sensitivity thresholds but also for monitoring the underground water contamination at testing. Usually the component is considered as detected if its concentration exceeds the value of its average concentration in the blank sample by more than three standard deviations ($C_i > \overline{C}_{i,b.s.} + 3 \cdot s_{i,b.s.}$).

Limit of quantification describes a level, above which the parameter value may be expressed with the assigned degree of reliability. The value of a component concentration is considered significative if it exceeds the value of its average concentration in a blank sample by more than 6-10 standard deviations ($C_i > \overline{C}_{i,b.s.} + 6 \cdot s_{i,b.s.}$).

Concentration values lower than DL do not mean either presence or absence of the analyte. At concentrations below DL the analyte is considered not detected. Concentration values higher than DL but lower than QL indicate that the analyte is present but its concentration is unknown. Such values are considered ***qualitative,*** their values are meaningless and are denoted as "traces" or "less than QL". Concentrations, which group around QL and reliably determine only the order of their values, are called ***semi-quantitative***. They are recorded as 10^n where n is the established order of the value. And at last, when the obtained data are much higher than QL values, study results may be considered ***quantitative***. Only in this case the analyte concentrations are expressed by specific values with the determined degree of reliability, and they may be compared between themselves.

2.2.4.2.2 Field Testing Quality Control

Factors affecting measurement quality parameters are possible to subdivide into three independent groups: natural, field and laboratory. They correspond with different liability levels.

Natural factors include mass-exchange with rocks, mixing in migration, contamination, etc. Their influence on the underground water quality is manifested before the testing and cannot be either eliminated or decreased. Moreover, the influence of natural factors is, as a rule, the object of hydrochemical study. Errors due to water properties and composition, natural micro-variations in points of testing, may be only random.

Laboratory factors are associated with the effect of the analyst on water samples in the process of their preparation, transfer in the device-analyzer and analysis. Their influence depends on the analysis technique, device calibration and quality, purity of chemicals, qualification of laboratory personnel, etc. Laboratory factors determine errors of water properties and composition analysis in a sample; these factors are within the competence of chemists-analysts and have inherent ***laboratory precision, laboratory bias***, but also DL and QL pertaining to techniques of water analysis in the sample.

These parameters must be known prior to field operations as they are the criterion for the selection of analytical techniques and also of the laboratory.

Field factors determine those underground water properties and composition changes, which occur at the time of the water becoming a sample. They are associated with the effect of climatic environment, conditions of water occurrence and extraction, its testing, preparation, safekeeping and transport. They determine the representativity of underground water samples. A sample is considered ***representative*** if properties and chemical composition of its water gives a complete idea of water properties and composition in reservoir conditions. It needs to be remembered that it is impossible to completely preserve formation water properties and composition at testing. But with careful testing it is possible to preserve its elemental composition. For this reason the main task in collecting representative samples is minimization of the loss of their own matter and also the introduction of foreign matter (contamination). The solution of these problems is within the competence of a hydrogeochemist.

Thus, the hydrogeochemist and analyst bear different liability for technical errors of testing results. The analyst is responsible for the reliability of measuring water properties and composition in a sample, hydrogeochemist – in the studied geologic object. Therefore, in hydrogeochemical studies it is impossible to limit quality control only to chemical analyses. In such studies it is necessary to take into account also field errors.

Field errors of the testing technique determine the boundaries of its suitability. For this reason any technique of hydrogeochemical testing must be a subject of continuous control, which may be performed as well in the process of field works, i.e., has operational nature, as after these works based on analysis of the collected samples. The purpose of such control is to determine the parameters of testing results quality and their matching with the values of acceptable error standards.

Any control is an experiment, which includes collection of control samples, their analysis and estimating the parameters of testing technique errors as a whole. At the stage of preparation of the analytical base, the contents of the field part of this experiment are determined, namely, the nature and number of needed control samples and procedure of their collection.

2.2.4.2.3 Control Samples

Field errors or ***sampling errors*** are directly associated with collection of water samples in the field and may occur due to incorrect planning of time and points, changes in water pressure and temperatures at its extraction, mass-exchange with equipment materials, chemicals, etc. Total field error is the sum of errors at four basic stages of testing preceding the analysis (Table 2.7), namely:

1. Selection of points and time of testing – $\Delta C_{i,x,y,z,t}$;
2. Affecting the water during its extraction on the surface – $\Delta C_{i,extr.}$;
3. Collection and preparation of samples – $\Delta C_{i,sampl}$;
4. Long safekeeping and transport into laboratory – $\Delta C_{i,tran.}$;

These errors are summed up in testing, and in the final analysis water in a sample may noticeably differ from the water in reservoir conditions.

1. Errors of selecting the testing points and time, as a regular rule, are caused by the nonuniformity of studied ground water. For this reason the cause of these errors usually is insufficient density and nonuniform distribution of testing points or time. The largest errors of such kind are associated with the objects located under conditions of drastically changing properties and composition, i.e., in boundary zones. In a single-time study these errors may be dominated by the systematic component, and in monitoring, by the random component. These errors are impossible to measure but they may be decreased by the literate and uniform distribution of testing points.
2. At the testing of deeply occurring waters the loss of dissolved matter is possible while lifting due to degassing or precipitation of salts in deposit. The consequence of this is the systematic negative error whose role may increase with depth. Random error at that will be associated with variations of this loss in the process of water extraction. These errors may be controlled only with the use of downhole hermetic samplers, which minimize loss of any matter.
3. At the water extraction from wells or shaft wells with special instruments, underground water touches and interacts with foreign materials of different composition and purity. The consequence may be the loss of the water portion of dissolved matter due to ion exchange and adsorption and its contamination. These processes occur prior to the sample collection and for this reason do not depend on their presence. They may have systematic nature. In order to decrease

Table 2.7 Sources of field errors in measurement of ground water properties and composition.

Source		**Causes**	**Controle**
Selection points and time of testing		Poor information about the nature of the object. Incorrect and nonuniform distribution of testing points or time.	Objective control unavailable
Water extraction on the surface	Changes in thermodynamic conditions	Lowering formation pressure and temperatures, contact with airом.	Uncontrolled systematic errors associated with loss of matter. Necessary to sample with downhole hermetic sampler.
	Interaction with materials of wells and instruments	Contamination, solution, ионный exchange in well, in shaft welle, in samplere, in bailing tubee etc.	Monitoring of composition and properties of water wells or shaft wells at their flushing. Blank instrumental samples of rinsing.
Samples transfer in vessel, its preparation and plugging.		Unsuitable material or polluted vessels for sample, filter, chemical, extractant, selection, etc.	Field blanks Repeated samples Dividable samples
Physicochemical processes in sample at the time of safekeeping and transport.		Freezing or insufficient hermeticity of vessel, precipitation of deposit, degassing, degradation of organics, etc.	Blank transported sample.

the influence of well (shaft well) materials and instruments, before testing the well is carefully flushed and the instruments are washed. With high liability level for errors, the instruments are rinsed with distilled water and the last rinse is taken for analysis ***(rinse or cleaning blank, equipment (rinsate) blanks***).

4. Large errors are possible directly at the collection of water samples for analysis. The lower the analyte concentration, the more difficult is its conservation and concentrating, and the higher is the risk of errors. However, at this stage samples are differentiated by their purpose and for this reason are distinguished by the preparation or concentrating methods. Due to this the values and nature of errors in collection of samples turn out to be dependent on their purpose. For this reason testing the quality control at this stage is conducted separately for each type of sample. With high liability level for errors, for sample collection error control, besides work samples, there may be up to 5 types of control samples collected: field blanks, repeated, split, spiked samples and samples for comparison.

Field blank samples are distilled water samples (see above) prepared with observance of all procedures of sampling for specific analyte. Field blank samples are necessary for estimating the sample contamination nature and value at the time of its preparation but also DL and QL value of the testing technique as a whole.

Replicate samples are work samples collected by the same method at one point. The measurement of standard deviation value in replicate samples of the same water point describes the testing technique precision as a whole, i.e., in consideration of within-the-laboratory precision. At monitoring studies these samples are collected almost simultaneously. In the case of a single-time study (hydrogeological survey, exploration operations, etc.), especially over a large territory, it is impossible to sample all points simultaneously. In this case collection of replicate samples is stretched over the entire period of field operations in order to take into account the influence of dynamic processes on the value of random errors. Without replicate samples it is impossible to estimate the precision of the testing technique as a whole. For this reason besides the work samples and their duplicates, at least one replicate should be collected. But economically beneficial is to collect 2 control replicate samples in one point, as with a small growth in cost the quality of the statistical estimate will be substantially higher. For a complete representativity estimate of the entire testing technique, it is necessary to have at least 20 replicate samples.

Split samples are work samples of large volume, which after preparation and before transfer to the laboratory are divided into samples of smaller volume. It is very important before dividing the sample to have it well stirred, homogenous, without outside deposit and suspension. Parts of split samples for the same parameter are sent in different laboratories for estimating reproducibility of the testing technique as a whole, and for finding possible inter-laboratory systematic errors.

Spiked samples are samples of water with the addition of a known amount of studied analyte. For this purpose special spiked-solutions are prepared with a specific concentration of analyte (Cu, Cd, Pb, Ag, etc.) whose content is balanced by such anions as a chloride, nitrate or acetate. Standard samples or certified mixes are also used as spiked-solutions. In studies of organic components, sometimes spiked-solutions are prepared, not of the analyte proper, but of its replacement, which is absent in the underground water but has the same properties as the studied component.

Spiked-samples are prepared in order to estimate the systematic error sign and value in the preparation of samples and their analysis. They enable both the control of sample contamination with foreign analyte and its loss at different stages of testing. At contamination, error values have positive readings and at the loss, negative. This control technique is especially important in studies of strongly polluted water when outside matter is present in such great amount that it plays the role of a cosolvent or extractant for the analyte. In these cases for the preparation of a spiked-sample of optimum quality it is necessary to have an idea of the contamination nature and anticipated analyte concentrations.

Comparison samples are uniform solutions with quite reliably determined parameters of the properties or composition, which are used for the calibration of devices and for estimate measurement technique qualities. Among comparison samples are distinguished standard samples, certified mixes and natural solutions whose composition is supported by different methods in different laboratories.

Standard samples are solutions with quite accurately measured and officially certified values of their chemical composition. Depending on the level of recognition (approval) and application domain they may be interstate, state, disciplinary or belonging to individual undertaking. In the RF, state standard solutions of both inorganic and organic matter are being issued, which are held in glass ampoules. In standard samples of metal solutions acid of 1-10% concentration is added to the solution.

Standard samples of organic compounds are, as a rule, solutions of high-purity organic matter in corresponding solvents. Standard samples are commonly used for the accuracy certification of techniques and parameters for measuring component contents in water and environmental objects.

Certified mix is the mix of two or more substances prepared by the standard technique and having certified values of analyte concentrations. Certified are not analyte concentrations proper but their determination under the calculation-experimental procedure of mix preparation. These mixes with the assigned analyte concentration are prepared by way of mixing several substances in compliance with a determined technique, as a rule, at the point of use. Certified mixes are developed and allowed for the application in the cases when the generation of a needed type standard sample is difficult due to low stability of used substance or economic inexpediency. As opposed to standard samples they must not be mass-produced. In particular, certified mixes are used in testing of pesticides, mycotoxins and other complex organic compounds.

Natural water solutions of sufficiently reliably determined composition also may be used as comparison samples. For instance, a US Geologic Survey proposed 6 types ***of standard reference samples***, which are the samples of special prepared surface water.

Comparison samples are used both for the preparation of spiked samples and independently in pure form for the determination of systematic errors. In the latter case comparison samples, similar to the blanks, are subjected to all testing procedures for samples of a given analyte. It is necessary for the comparison samples to maintain stability during the conduct of the experiment and to contain sufficient amount of the analyte to be determined. For a reliable control at least three comparison samples are recommended.

2.2.4.2.4 Selection of Error Control Technique

It would appear, the higher the accuracy and sensitivity of a hydrochemical study, the more reliable is the derived material and higher is the quality of the works. However, improved accuracy and sensitivity of the testing technique increases the cost of work. This is associated with an increase in the number and cost of analyses for both work and control samples. Total number of control samples may reach 15% or more of the number of work samples.

The number and nature of control samples, methods of their collection and analysis to a significant extent depend on the objectives and tasks of a hydrogeochemical study; more accurately, they depend on the level of liability for errors. In some cases it is necessary to establish only the nature of unknown component, in some other cases the presence of a certain component. The task assignment here is not the amount but reliable identification of the studied component. Such task is often to find dangerous organic pollutants or pathogen microorganisms. For its solution a reliable qualitative analysis is sufficient.

At the preliminary stages of a hydrochemical study it is often sufficient to estimate only approximately the concentration value for substantiation of more reliable ground water testing techniques. In such a case a semi-quantitative analysis is sufficient.

Quality of measuring the contents of major macro-components (global analysis) for a general hydrochemical description of ground waters is usually not restricted by a margin of error and is not controlled. However, if it is required to determine the analyte's concentration distribution in time and space, testing and analysis methods must guarantee the margin of measurement error by one order of the magnitude lower than the background analyte content in ground waters, and absolute testing error at least 3 times less than the range of the studied variations. Error measurements of the content of many mineral, organic or even gas components is often restricted by a specific standard of errors whose value is set directly by the client or by regulatory documents.

Especially great is the liability for errors in hydrogeochemical studies in solving tasks associated with ground water contamination. Missing a dangerous substance or underestimating its content in water may have deadly consqeuences. On the contrary, erroneous finding or overvaluation of dangerous substance content may cause extra costs for their control and elimination. Especially great is the liability for a testing error when the derived results are used for expert evaluation of ecological liability, especially if the ecological crime was committed. In this case a court judgment may depend on the determination accuracy of concentrations or isotope composition. Correlation between the required reliability and liability level is provided in Table 2.8.

2.2.4.3 *Selection of Chemical Analysis Technique*

Selection of chemical analysis technique depends, first of all, on the selection of studied analytes, on the acceptable measurement errors and on points of their conduct (in the field or in a special laboratory).

Selection of studied analytes depends on task assignment, is determined at the selection of necessary property and composition parameters and serves as a basis for the selection of chemical analysis technique. Acceptable

Table 2.8 Control of hydrochemical testing errors vs. level of liability for error.

Level of liability	Reliability of measured amount	Objectives of hydrochemical study	Control sample
1	Qualitative analysis	Determination of the presence of rare dangerous matter or pathogen microorganisms	Dividable
2	Semi-quantitative analysis	Measurement of heavy metals, organic compounds content in reconnaissance and exploratory operations	Repeated, dividable
3	Quantitative without reliability control	Measurement of content and distribution of major macro-components in ground water general composition	Repeated, dividable
4	Quantitative with control only of random errors	Measurement of parameters under the same technique	Repeated, dividable, blank
5	Quantitative with control of random and systematic errors	Quality of drinking water, comparing waters quality at different objects, measurement of parameters under different techniques	Repeated, dividable, blank, spiked-sample
6	Quantitative or qualitative very high accuracy	Measurement of parameters used in expert evaluation of ecological liability	Repeated, dividable, blank, spiked-sample, standard

measurement errors also depend on such assignment; they are determined by the level of liability for study result errors and are established before selecting techniques of the chemical analysis.

At last, measurement of water property and composition parameters may be conducted either in the field (on the surface or directly at the depth of testing) or in a stationary laboratory, to which the water sample is transported.

In the field, ground waters are usually studied on the surface, directly at the water show. They may be numerously repeated, which allows for the operative control of measurement quality. For this reason easier requirements are imposed on field measurement methods. Greater significance is placed on simplicity and performance of these techniques, portability, transportability and stability of their tools. Field methods are oriented mostly on the measurement of most accessible or least stable parameters (organoleptic properties, temperature, electric conductivity, radioactivity, gas-saturation, NH_4^+, NO_2^-, NO_3, pH, Eh etc.). Field studies often are conducted for a purpose of optimizing testing and analysis techniques in stationary laboratories.

Complex and subtle ground water studies with the application of bulky stationary equipment are conducted in special laboratories, which may be thousands of kilometers away from a study object. In these cases the stability of sample properties and composition during the time of their transport and safekeeping is of great significance. Besides, the water analysis is conducted by chemist-analysts who, as a rule, do not have preconceptions either about the task to be solved or about particulars in the composition of analyzed waters, or about conditions of their collection and safekeeping. Therefore, a significant part of the liability for the correct selection of analytical technique and laboratory, for a competent estimate of the obtained results, is on the hydrogeochemist. For this reason while selecting analytical techniques, it is necessary to have at least a general idea of their nature, intent and accuracy.

Currently the analytical chemistry offers a long series of ground water analysis techniques. They all include two main operations: 1. Acting on the analyzed components (analytes) in water composition in order to force them to reveal themselves by some indications; and 2. Determining their content by way of recording these indications. Affecting analytes may be different:

- forcing them to precipitate or form a suspension;
- attributing to them a specific color;
- exciting their atoms or molecules by electromagnetic or other radiation;
- dividing them by size, charge or sorption capacities, etc.

These effects to various extent destroy the integrity of solution components. Milder analytical methods (chemical, electrochemical, chromatographic and also molecular spectroscopy and capillary electrophoresis) preserve the strongest compounds in water composition, i.e., ions, molecules or even some complex compounds. But with these analytical methods the analytes are usually brought preliminarily to a single analyzed form. These methods determine the ***component composition*** of ground

Table 2.9 Comparison of ground water chemical analysis techniques

Methods	Type of technique	Determined components	Sensitivity, $g \cdot l^{-1}$	Global error, %
Molecular analysis				
Chemical	Gravimetric	ions, I_2, CO_2, organic acids, petroleum products	0.0001	0.5–1
	Titrimetric	ions, O_2, CO_2, H_2S, COD, organic nitrogen, organic acids	10^{-3}–10^{-4}	0,5–1
	Gas volumetric	Hydrocarbons, N2, O_2, CO_2, Ar, He, etc.		
Electrochemical	Potentiometry: ion-selective electrodes	Ions	See Table II.10.	
	Volt-amperometry: Volt-amperogram	Ions	10^{-3}–10^{-7}	1–2
	Stripping volt-amperometry	Ions		
	Capillary electrophoresis	Ions		
Chromatographic	Liquid ion-exchange	ions, organic acids	0.0001	–
	High-efficiency liquid	Organic components		
	Gas-liquid	Gas and organic components		
	Gas-adsorption	Gas and organic components		

(*Continued*)

Methods	Type of technique	Determined components	Sensitivity, $g \cdot l^{-1}$	Global error, %
Spectrometrical	Photo-colorimetric (optical, UV)	ions, organic compounds	10^{-2}–10^{-3}	1–5
	Specter-photometric (optical, UV)	ions, organic compounds, petroleum products	10^{-5}–10^{-7}	5–10
	Fluorimetry (luminiscence)	Ions, petroleum products, surfactants, organic compounds	10^{-3}–10^{-7}	0.5–10
	IR spectrometry	Organic compounds, petroleum products		
	Raman scattering spectrometry	Organic compounds		
Mass spectrometric	Analytical thermal ionization	ions, hydrocarbons, N_2, O2, CO_2, Ar, He, etc.		
Elemental spectroscopic analysis				
Atomic absorption (AAS)	Flaming		See Table II.11.	
	Electro-thermal			
Atomic emission (AES)	Flaming photometry			
	Induction-associated plasma			
Atomic fluorescence (AFS)	Flaming fluorescence			
	Electro-thermal fluorescence			
Mass spectrometric	Elemental induction-associated plasma			

water. Other methods, which destroy compounds to free atoms, determine ***elemental*** or even ***isotope composition***. For this purpose the water samples are boiled down and the precipitate is brought in the plasma state.

The content of identified analytes is determined from such attributes as weight, volume, usage of chemicals, color or spectrum and their intensity (photometry, emission and sorption spectrometry), etc. Depending on this, all analytical methods used in studying ground waters are subdivided, first of all, into chemical and instrumental. The former are conducted through the use of chemical reactions in the solution and measurement of the weight (gravimetry), volume (volumetry) or usage (tittering) of chemicals. The latter ones are performed through the measurement of analyte's physical properties with electrochemical, optical, radiochemical and other devices. Among instrumental techniques for analysis of water composition are commonly used electrochemical, chromatographic, spectrometric, mass-spectrometric, etc. In recent times a new instrumental technique, capillary electrophoresis, has been actively introduced, which has the attributes of both chromatographic and electrochemical analyses. At last, there is a unique technique of elemental activation analysis relatively rarely used in hydrogeochemistry, with which the matter is irradiated by nuclear particles (thermal or fast neutrons, protons, deuterons, α-particles, etc.) or γ-rays, then from half-life periods the type and mass number of formed radionuclides is determined. Following is a description of the most common techniques of water composition analysis.

2.2.4.3.1 Chemical Analytical Methods

Chemical methods are used directly in the sample with special chemicals. They take the analyte from solution, and its content is determined by the weight or volume of the obtained products. Chemical methods are distinguished by relatively mild effect on the analyte and enable the determination of the content of ions and larger and more complex molecules. They are highly accurate. Major application domain of chemical techniques is precision determination of high and medium concentrations of mineral components. These methods, however, are labor-intensive and gradually give way to instrumental. In hydrogeochemistry the most popular among them are gravimetric, titrimetric and gas volumetric methods.

a. <u>Gravimetric</u> (weight) methods include extraction of an analyte from the solution with the help of a chemical and subsequent determination of its weight. Depending on the manner of extraction methods of precipitation, distillation and separation are distinguished. In the first case the component is converted to deposit, which is then rinsed, dried up or even annealed and weighed. The distillation technique is applicable to volatile compounds. It boils down to the separation of analyzed components in the form of gaseous matter and the determination of either directly its mass (direct technique) or mass of the chemical's residue (indirect technique). Under the separation the analyte is deposited on the electrode by electrolysis and then the electrode with the deposit is weighed. Gravimetry was in common use until the arrival of spectral and chromatographic methods. It is distinct in relatively high accuracy. Currently it remains a reference, methodological base for the development and certification of other techniques. Global relative error of the gravimetry is 0.1–0.2%. In the analysis of complex composition samples the error increases to several percent due to imperfection of analyte-dividing and separating techniques. The selectivity of gravimetric techniques is not high. Besides, a large drawback of the gravimetry is low speed. Gravimetric methods are applied for the determination of Mg^{2+}, Na^{+}, SiO_4^{4+}, SO_4^{2-}, I_2, CO_2, total content of petroleum products, humic acid and fats. Gravimetric methods require very accurate weighing under stationary conditions and practically are not suitable for field use.

b. Titrimetric methods are based on the usage determination of a chemical (titrant) expended for neutralizing the determined component in solution. The moment of equivalency is usually established by the color change of a specially selected indicator. Depending on the type of utilized reactions are distinguished titterings: acid-base (neutralizing), oxidation-reduction (iodimetry, iodimetry, permanganatemetry, etc.), complexometry and precipitation methods.
Tittering still remains the most common chemical technique of global water analysis. It is used to determine the content of Ca^{2+}, Mg^{2+}, Cl^{-}, SO_4^{2-}, CO_3^{2-} and HCO_3^{-}, Fe^{3+}, Cu^{2+},B, O_2 (Winkler's technique), CO_2 and H_2S, fixed nitrogen (Kjeldahl's technique), organic acids, total hardness, acidity, alkalinity, COD, etc. Major advantages of titrimetry are simplicity, determination speed and the possibility of automation. Its lower limit of detection is 10^{-3}–10^{-4} %; the error is 0.5–1.5%. These methods are often used in field conditions.

c. Volumetric techniques are used as a rule for the composition analysis of natural gases whose components have exceptionally low chemical activity. The technique boils down to sequential removal of components and measuring the volume of the remainder. Components are removed with the help of chemical reactions as well as sorption or liquefaction, their content is expressed in fractions of the volume. The volumetric technique is complex, requires high vacuum and is very labor-intensive. It is applicable only under stationary laboratory conditions. It was practically displaced by gas-chromatography technique. But it is convenient when it is necessary to identify individual gas components in pure form for isotope analysis. This technique is utilized for the determination of relative volume of most components in the natural gas, including N_2, Ar and He.

Chemical methods are the oldest, relatively simple and do not require analysts of very high qualification. They are often called classical, and among them gravimetric and titrimetric are attributed to "wet analyses". Without chemical methods is not possible to determine total alkalinity, hardness and SO_4^{2-}. In most cases Cl^- is also determined by chemical methods. But it is necessary to take into account that these methods may not notice stable complex compounds of the analyzed components and thereby understate their concentrations. For this reason they have unstable accuracy and currently are almost completely displaced by modern instrumental physicochemical methods. Among the latter, the most significant in hydrogeochemistry became the electrochemical, chromatographic and photometric methods.

2.2.4.3.2 Electrochemical Analysis Methods

Electrochemical analysis is based on the capability of solutions to affect parameters of closed electrochemical chain. Depending on the measured parameters they are subdivided into five basic groups: ***potentiometry***, ***voltammetry***, ***conductometry***, ***coulometry*** and ***dielectrometry***. For ground water analysis only the first three techniques are used. All electrochemical methods use electrodes and differ mostly in their nature and measured parameters. The analytical system of these techniques includes, as a minimum, two electrodes – ***indicator*** (measuring electrode) and ***standard*** (comparison electrode). The indicator electrode must rapidly and reversibly react to changes in the concentrations of a measured ion. The standard electrode has permanent, unchangeable in the course of an analysis, potential. The sensor joining both electrodes in one frame is often called ***combination electrode***. For the implementation of an analytical task through electrochemical methods, the correct selection of the system of indicator and standard electrodes is most important. In hydrochemistry potentiometry and voltammetry are usually applied.

a. ***Potentiometry*** includes methods based on measurements of electromotive force of reversible electrochemical chains, when the potential of the working electrode is close to the equilibrium value. In potentiometry, mostly selective (membrane) indicator electrodes are applied, which selectively react to the ion (or group of ions) to be determined and change their potential depending on its activity (concentrations) according to the Nernst's equation. For this reason the potentiometric analysis covers only the matter in the ionized state. Their potential value is measured relative to the standard electrode whose potential does not depend on the composition of the solution and is tentatively assumed equal to zero. Potentiometry includes two large and equitable sections - ***ionometry*** and ***redoxmetry***. The former deals with measurement of ion concentrations (activities) and the latter operates with redox-potentials characterizing the concentration and the ratio of oxidized and reduced forms. The measuring electrodes for ionometry are usually subdivided in two groups: pH-***electrodes***, i.e., those selective to the hydrogen ion, and ***ion-selective electrodes***, i.e., those selective to other ions. Beside electrodes, a device is needed for measurement potentials, which in the first case is called pH-meter and in the second, ionometer (Figure 2.12).

   The techniques include direct potentiometry and potentiometric tittering. The most accessible in the field is ***direct potentiometry*** using ion-selective electrodes. It is usually called simply ***ionometry***. This is the only technique determining not the concentrations but directly ion activity in solutions. Similar methods are sometimes called low concentrations technique as their concentration measurement errors increase with growth in salinity.

   Depending on the sensitive membrane material, the modern ion-selective electrodes are glass, crystalline or film (Figure 2.13). ***Glass ion-selective electrodes*** have best

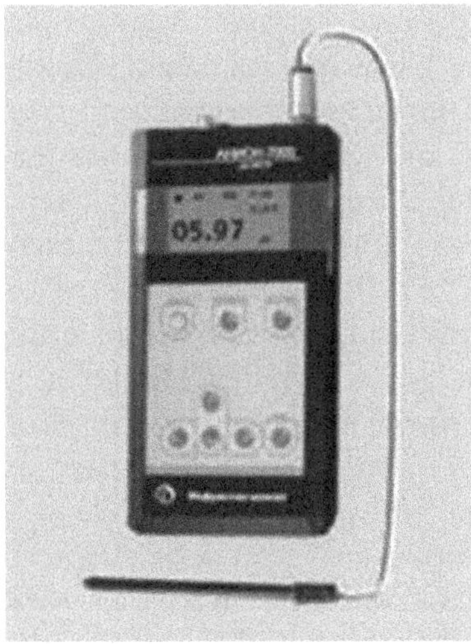

Figure 2.12 Ionometer **Anion-7010** intended for measurement of ions activity (pX/pH) and ion molar and mass concentrations, E.M.F. of electrode systems, oxidation-reduction potential (Eh) and temperature of water medium.

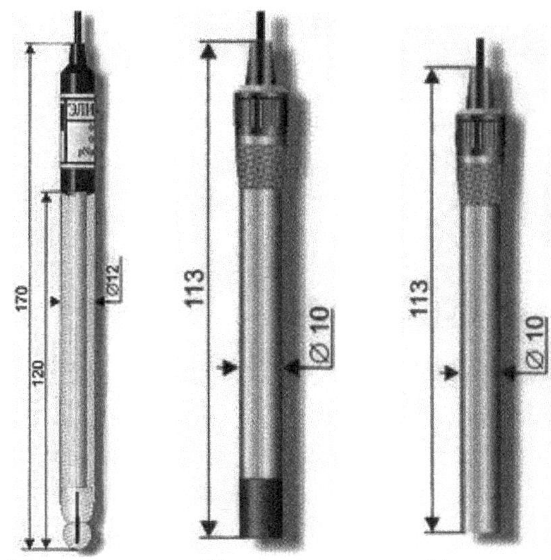

Figure 2.13 Ion-selective electrodes (series ELIS): a- glass, b - with crystalline membrane, c - with PVC membrane.

exploitation parameters. Application of glass electrodes is impossible only in solutions containing fluoric acid or its salts. ***Crystal-membrane electrodes*** are manufactured with plastic bodies. For this reason they are not recommended for use in solutions containing strong oxidants and organic solvents, which destroy or dissolve the body material polyvinylchloride (PVC). The electrode sensitivity of this type (except fluorine) may be restored by cleaning the surface with fine sandpaper. And the electrodes with film membrane (*PVC-membrane electrodes*) demand careful treatment. Their membranes, like their bodies, are manufactured from PVC, but with the addition of a special matter - carriers. These membranes cannot be rubbed or mechanically cleaned. Working with them excepts the presence of a matter dissolving or destroying PVC.

Reliability of potentiometry results strongly depends on correctness of the electrode selection. First of all, it is necessary to assure their suitability. In case of doubt it is recommended to ask experts-analysts and developers of the technique. Selecting electrodes, it is important to take into account the conditions and range, in which they reliably determine the analyte concentration.

Reliability of analyses is affected by pH, temperature and presence of outside ions (Table 2.10). In some cases pH lowering with acid is acceptable. The temperature operating range of pH-electrodes is 0 to 150°C. For ion-selective electrodes this range is much narrower. If it is not possible to find electrodes with needed temperature range, the analyzed solution is heated or cooled.

Operating concentration ranges determined by selective electrodes are listed in Table I39. Concentrations higher than the operating range may be lowered by the addition of distilled water. The analysis is more complex at concentrations lower than this range. Work with indicator selective electrodes requires their mandatory preliminarily calibration. The potentiometry accuracy depends, first of all, on the type and qualities of the measuring electrode. For ion-selective electrodes it is total relative error value at 4% for single-charged ions and 8% for double-charged. It should be remembered that potentiometry determines directly ion activity whose deviation from the concentration grows with

Table 2.10 Properties of ion-selective electrodes of different purpose.

Ion	Detection limit, mg/l	Operating range, mole·l^{-1}	Operating range, pH	Obstacle ions
Film electrodes				
K^+	0.40	10^{-5} – 10^{-1}	1 – 10	Na^+, NH_4^+, Ca^{2+}
NH_4^+	0.20	10^{-5} – 10,0	1 – 13	Na+, K +, Ca^{2+}
Ba^{2+}	1.40	10^{-5} – 10^{-1}	3 – 10	Na+, K +, Mg^{2+}
Ca^{2+}	2.30	2×10^{-5} – 10^{-1}	4,5 – 10	Na+, K^-, Mg^{2+}
Ca^{2+}, Mg^{2+}	0.01 meq·l^{-1}	10^{-5} – 1,0	5 – 10	Na^+, NH^{4+}, Ba^{2+}
ClO_4^-	1.00	10^{-5} – 1,0	0 – 12	Cl^-, NO_3^-, CNS^-
NO_3^-	1.30	2×10^{-5} – 2×10^{-1}	1 – 10	Cl^-, NO_2^-, CO_3^{2-}
CO_3^{2-}	0.60	10^{-5} – 3×10^{-2}	6 – 9	CH_3COO^-, SO_4^{2-}, CNS^-
A surfactant	3.00	2×10^{-5} – 8×10^{-3}	2 – 8	
Crystalline electrodes				
Ag^+	0.10	10^{-6} – 1,0	1 – 9	Hg^{2+}
Cd^{2+}	0.10	10^{-6} – 10^{-1}	3 – 7	Ag^+, Hg^{2+},Pb^{2+}
Cu^{2+}	0.06	10^{-6} – 1,0	2 – 7	Ag^+, Hg^{2+}
Hg^{2+}	0.20	10^{-6} – 10^{-1}	0 – 2	Ag^+
Rh^{2+}	0.20	10^{-6} – 10^{-1}	3 – 7	Ag^+, Hg^{2+}, Cd^{2+}
Br^-	0.40	5×10^{-6} – 1,0	1 – 13	I^-, S^{2-}
Cl^-	1.75	5×10^{-6} – 1,0	1 – 13	Br^-, I^-,
CN^-	0.03	10^{-6} – 10^{-2}	11 – 13	Br^-, S^{2-}
CNS^-	0.60	10^{-5} – 10^{-1}	2 – 12	Cl^-, Br^-,I^-
F^-	0.02	10^{-6} – 1,0	4 – 7	
I^-	0.06	5×10^{-7} – 1,0	1 – 13	S^{2-}
S^{2-}	0.03	10^{-6} – 1,0	13 – 14	Hg^{2+}
Chalcogenide glass electrodes				
Ag^+	0.010	10^{-7} – 1,0	1 – 7	
Ti^+	0.200	10^{-6} – 10^{-1}	2 – 7	Ag^+, Hg^{2+}
Cu^{2+}	0.006	10^{-7} – 1,0	0 – 7	Ag^+, Hg^{2+}
Pb^{2+}	0.020	10^{-7} – 1,0	2 – 7	Ag^+, Hg^{2+}
Cd^{2+}	0.010	10^{-7} – 1,0	1 – 7	Ag^+, Hg^{2+}
Hg^{2+}	0.020	10^{-7} – 1,0	0 – 2	
Fe^{3+}	0.060	10^{-6} – 10^{-1}	1 – 2,5	Ag^+, Hg^{2+}
Cr^{6+}	0.006	10^{-7} – 10^{-3}	0 – 2	Ag^+, Hg^{2+}
S^{2-}	0.003	10^{-7} – 10^{-1}	3 – 9	

increasing salinity, and that eventually, with the depletion of electrode resource, measurement errors increase.

The introduction of ion-selective electrodes enabled significant broadening of their practical use for analytical control of pH, Eh values and concentrations of O_2, Na^+, Ca^{2+}, Mg^{2+}, NH_4^+, NO_3^-, Cl^-, F^-, etc., directly in the field conditions. Glass electrodes are currently best pH-electrodes as they are practically insensitive to Eh and highly selective within broad range of pH (1 to 12.5). Electrodes allow performing measurements of medium's acidity with the accuracy of up to ±0.03 pH units. For ionometry a broad range of measurements within $1-10^{-6}$ mole·l^{-1}, and in some cases to 10^{-8} mole·l^{-1}, is typical.

b. Volt-amperometry (***voltammetry***) is based on measuring correlation between the polarizing current on the indicator electrode and changes in voltage supplied to the solution. This is a non-equilibrium technique of electrochemical analysis based on electrolytic processes of individual ions. At the running of electric current and ion electrolysis the indicator electrode is polarized and its potential noticeably deviates from the equilibrium values. For magnification of polarization special indicator micro-electrodes are used, which have a much smaller surface area than standard electrodes. They are manufactured of mercury, platinum and electro-conductive carbon materials. The electric current density on indicator micro-electrodes are four orders of magnitude greater than on standard ones. The derived correlation between the current and the voltage, i.e., volt-ampere curve, is called ***voltammogram***. It has peaks whose position describes the electrolyzed analyte, and their height is proportionate to its concentration in the solution.

In terms of the variety of techniques, voltammetry is the most numerous and commonly used group among electrochemical analysis techniques for the determination of compounds in solutions. More often than others the two most sensitive techniques are used: polarography and stripping voltammetry.

The ***polarography*** is a voltammetry using as the indicator electrode the so-called ***dripping mercury electrode*** (Figure 2.14). On this electrode, drops of mercury flow out of the capillary under pressure, providing thereby permanent renewal of its surface and ideal polarizability within a

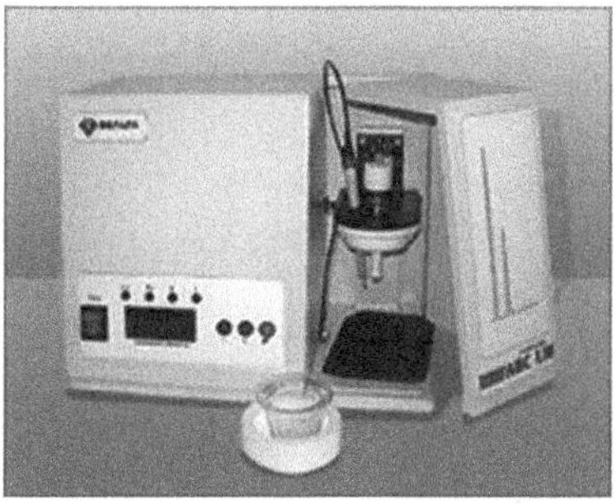

Figure 2.14 Voltammetric analyzer (polarograph) ABC 1.1, intended for the determination of heavy metals (Cu, Pb, Cd, Zn, Bi, Ni, Ag, I, As, Fe, Sb), in potable, natural and waste waters.

wide interval of potentials. Voltammograms derived by this technique are often called ***polarograms***.

Stripping voltammetry includes preliminary electrolysis accumulation of analyzed matter on the indicator electrode at a constant potential and its subsequent electrochemical dissolution due to linear changes in the potential. The preliminary accumulation is conducted both by cathode polarization on the stationary electrode (for determination of cations) with subsequent anode dissolution, and by anode polarization (for the determination of anions and organic compounds) with subsequent cathode dissolution.

Polarography and stripping voltammetry guarantee high analysis sensitivity and speed. They enable the determination of both inorganic and organic component contents in several (up to 4–5) at once without their division. The stripping voltammetry, compared with the regular polarography, allows increasing the sensitivity by 2–3 orders of the magnitude. The lower limit of determined concentrations in voltammetry methods is 10^{-5}–10^{-6} mole·l^{-1}. In stripping voltammetry this limit goes down to 10^{-9}–10^{-11} mole·l^{-1}. Currently stripping voltammetry is mostly used for the determination of metal content (Ag, As, Cr, Cu, Pb, Cd, Ni, Co, Sb, Bi, Mn, Se, Zn).

c. Conductometry is based on measuring the electric conductivity of natural waters. Electric conductivity of diluted electrolyte solutions depends on the number (concentration) and charge of the ions. Conductometry includes direct (used, for instance, in salt meters) and indirect (for instance, in gas

Figure 2.15 System of the capillary electrophoresis "KAPEL®-104T," manufactured by "Lumex" - a device of high-efficiency capillary electrophoresis with UV photometric sensor.

analysis) analytic methods with application of the direct or alternate current (low and high frequency) and also chrono-conductometry, low-frequency and high-frequency tittering. In hydrochemistry this technique is used mostly for the determination of total salinity with portable conductomers (Figure 1.17).

d. Capillary electrophoresis is an analytical technique combining electrochemistry and chromatography (Figure 2.15). The substance of the technique is in dividing ions in a quartz capillary (diameter 50–75 μm) by the action of a high-voltage field (up to 30 kV). For this a small volume (about 2 nl) of the sample solution is introduced from the side of the anode into the buffer work solution of the capillary. Under the influence of a difference in potentials in capillary, the electro-osmosis flow (EOF) of the solvent forms, in which due to a high voltage, cations and anions migrate in the opposite directions. The ion migration velocity is the sum of EOF velocity and electro-migration velocity. As a result, at the capillary output first cations show up, then neutral compounds and the last are anions. The appearance of the components at the output is recorded by a UV photometric detector directly in the capillary. The outcome is a sequence of peaks called ***electropherogram***.

The technique of capillary electrophoresis is a high-speed, high accuracy instrumental technique, which may find broad application in analytical practice for the separation and quantitative determination of components in complex mixes. It is applied for quality control of natural, potable and waste water and allows the determination of Cl^-, SO_4^{2-}, NO_3^-, NO_2^-, F^-, PO_4^{3-}, Br^-, J^- and other inorganic and organic anions; and also NH_4^+, Na^+, K^+, Mg^{2+}, Li^+, Sr^{2+}, Ba^{2+}, Ca^{2+}, Mn^{2+} and other inorganic and organic cations. Currently a number of ion analyzers exists based on this technique. In particular this technique determines the content of dissolved acetate-ions (CH_3COO^-). Traditionally, this technique is compared with high-efficiency liquid chromatography (HELC), as in both cases dividing/separation occurs within limited space. At the direct photometric detecting the lower limit of detection (DL) is within 10^{-7}–10^{-4} mole·l^{-1}.

2.2.4.3.3 Chromatographic Analysis Methods

Chromatographic methods include physicochemical dividing of the solution components with the participation of heterogeneous interaction of two media: immobile – sorbent and mobile – eluent. Depending on the nature of the heterogeneous interaction adsorption, distributive, ion–exchange and other chromatographies are distinguished. The adsorption one is based on distinction in sorbilg capacity of adsorbent components; the distributive, on their different solubility in the immovable phase and eluent; the ion–exchange, on the difference in the constants of ion–exchange equilibrium, etc. In accordance with aggregate state of the eluent, ***liquid*** and ***gas*** chromatography are distinguished. In liquid chromatography a liquid is used as eluent. Depending on the aggregate state of the sorbent it may be solid–liquid or liquid–liquid. In a hydrochemical analysis, most commonly used is solid–liquid chromatography, which, in its turn, is subdivided by the nature of heterogeneous processes into ion–exchange and adsorption ones. Gas chromatography uses as the eluent a gas–carrier (helium, nitrogen, argon or other pure gas). Depending on the aggregate state of the immovable medium it may be ***gas–adsorption*** (adsorbent solid) and ***gas–liquid*** (adsorbent liquid).

These methods are implemented through the devices called ***chromatographs***. In these devices the eluent of the known composition is run through the chromatographic column with sorbent at a constant velocity. A precisely measured dose of a sample is introduced in this flow. Sample components, due to heterogeneous processes, migrate through the column at different velocities and thus are separated. As a result the eluent carries components out from the column in determined sequence. A detector at the end of the column measures the concentration change in time. The derived curve is called a ***chromatogram***. From its parameters concentrations of components are determined. If the component nature is unknown, it is determined after dividing the analyzed mix by different chemical, physical and physicochemical methods.

Figure 2.16 Ion chromatograph "Steier" intended for qualitative and quantitative analysis of inorganic ions F^-, Cl^-, NO_3^-, NO_2^-, Br^-, SO_4^{2-}, PO_4^{3-}, Li^+, Na^+, NH_4^+, K^+, Mg^{2+}, Ca^{2+}, Sr^+, etc., and also organic ions in water solutions and waters of diverse origin (natural, waste, potable, including bottled).

The major advantage of chromatographic techniques is the capability of dividing substances with close properties. For this reason chromatography is used for analysis as well as for preparatory dividing of mixes. Among numerous varieties of these techniques in hydrochemistry more often than others ***ion-exchange*, *liquid adsorption, gas-liquid and gas*** chromatographies are used.

a. Liquid ion-exchange chromatography uses as eluent a water solution of strong electrolytes with high electric conductivity (Figure 2.18). It is based on ion-exchange dividing of organic or inorganic ions. The sorbents are special ***ion-exchange resins*** (*ionites*), which are insoluble in water polymers or silica jells with ionogenic groups. Ion-exchange resins for cations (Na^+, K^+, Mg^{2+}, Ca^{2+}, Fe^{3+}, Cu^{2+}, NH_4^+, etc.) contain ionogenic groups like $-SO_3H^+$, $-COOH$, $-OH$, $-PO(OH)_2$ and are called ***cation-exchange resins*** (*cationites*) (cation-exchangers) and for dividing anions (F^-, Cl^-, NO_3^-, SO_4^{2-}, PO_4^{3-} etc.) contain ionogenic groups like $-N^+(CH_3)_3Cl^-$ or $-N^+(CH_3)_3OH^-$ and are called ***anion-exchange resins*** (*anionites*) (anion-exchangers). Also in existence are amphiprotic Ion-exchange resins (amphonites) capable of exchanging both with cations and anions.

Most commonly used is *high-performance ion-exchange chromatography* (HEIC) called sometimes simply ***ion chromatography***. Its efficiency is achieved by using additional suppressing column with subsequent conductometric detection (dual-column chromatograph). The eluent with already divided ions is fed into the suppressing column where its electric conductivity hindering the analysis is suppressed. The technique is distinct in high selectivity, low detection limits (to 10^{-3} μg·ml^{-1} and with the application of preliminarily concentrating – to 10^{-5}–10^{-6} μg·ml^{-1}) and speed (it is possible to determine up to 20 ions in 20 minutes). This technique operates in a broad range of determined concentrations (0.01 μg·ml^{-1}– 100 мg·ml^{-1}), low sample usage (0.1–0.5 ml) and easiness of sample preparation for the analysis. Ion–chromatography is possible to apply for simultaneous determination of inorganic and organic ions. It is also possible to determine cations of alkali and alkali–earth metals and some amines.

Ion chromatography is currently used mostly for analysing inorganic anion (F^-, Cl^-, Br^-, PO_4^{3-}, SO_4^{2-}, NO_3^-, NO_2^-, etc.) content.

b. *Liquid adsorption chromatography* is used as an eluent for easily volatile organic solvents (for instance, hydrocarbons, ethers, alcohols) and as an immobile medium for silica jells, alum gels, and porous glass with particle size of several micro–meters. For improving its efficiency programmed change in time of the eluent properties by immixing solvents of varying polarity is sometimes used. Elevated pressure, up to 50 mn·m^{-2} (500 kgf·cm^{-2}) allows running the eluent through sorbents with small sized particles (5–30 μm) and shrinking the analysis time from 2–3 hours to a few minutes. Such liquid adsorption chromatography is called ***high performance liquid chromatography*** (HPLC).

This technique is used, mostly, for dividing organic compounds. It allows researchers to determine the contents of saturated hydrocarbons, aromatic polycyclic hydrocarbons (anthracene, etc.), polar aromatics (phenols and phenol–derivatives), heterocyclics (tiophens, benztiophens, etc.), surface–active matter, etc. For the determination of these compounds the analysis is often conducted in combination with other physicochemical and physical methods, for instance with mass–spectrometry, IR–, UV–spectroscopy,

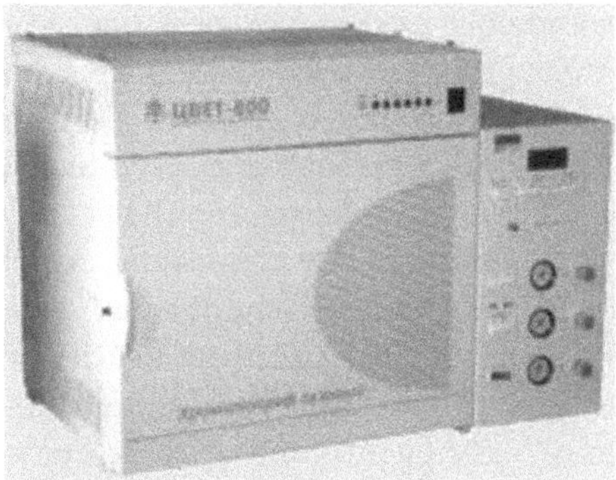

Figure 2.17 Gas–chromatograph Colo^{-4} to $1{\cdot}10^{-6}$ without concentrating, and $1{\cdot}10^{-6}$ to $1{\cdot}10^{-10}$% with concentrating.

etc. The use of high–sensitivity sensors allows working with very small amounts of matter (10^{-11}–10^{-9} g). Computers are used for deciphering of chromatograms and selection of the experiment conditions.

c. Gas–adsorption chromatography uses gas (helium, nitrogen, hydrogen, argon, air) as the eluent. The gas penetrates through fine-pored adsorbents (zeolites, porous polymers) when analyzing gas components or through macroporous adsorbents (carbonous, mineral and polymer) when analyzing volatile organic components (Figure 2.17). Gas-adsorption chromatography is commonly used for analysis of natural gas composition. As CO_2, easily adsorbed by molecular sieves, may hinder gas analysis, it is sometimes preliminarily removed.
d. Gas–liquid chromatography differs from the gas–adsorption chromatography in that the sorbent is prepared by spreading a liquid (high–boiling hydrocarbons, esters, siloxanes, etc.) over the surface of a solid carrier as a film several micro–meters thick. The difference between these techniques is in the nature of the detectors used.

Gas chromatography is applied for dividing and analysis of natural gases and organic matter that is easily transferrable to a vaporized state without decomposition. Gas chromatography is especially effective in dividing compounds of the same class (hydrocarbons, organic acids, alcohols, etc.). For that, several columns with sequentially varying selective properties are often used. It is used for dividing gasolines, kerosenes, oils (lubricants), pesticides and even metals, converting them into volatile complex compounds. Gas chromatography requires only a few cubic centimeters or milligrams of the analyzed matter, and the duration of analysis is several seconds to several hours. This chromatography is commonly used for content analysis of numerous organic compounds (acetaldehyde, acetonitrile, benzene butyl acetate, butyl alcohol, hexane, decane, isopropyl, isopropyl alcohol, crotonic aldehyde, buturic aldehyde, α–methylstyrene, methylvinyl pyridine, methylethyl pyridine, styrole, toluene, phenol, ethylacetate, and 2ethyl–1hexanole). In particular, with its help chlororganic pesticides, phosphorus organic and sym–triazine herbicides are analyzed.

Gas–liquid chromatography is widely used for analysis of organic compounds (acetone, benzene, methanol, toluene, xylenes, styrole, ethylbenzene, chloroform, phenol, carbon tetrachloride, volatile halogen organic compounds, etc.).

Thousands of chromatographic analysis techniques currently exist. Their selection is determined by the composition of the analyzed sample, by the analytical task, and by the available equipment. Accuracy and detection limits of chromatographic analyses depend on their equipment. Under favorable conditions the detection limit may reach $3 \cdot 10^{-5}$%.

2.2.4.3.4 Spectrometry Analysis Methods

Spectrometry methods are based on a capability of atoms and molecules in the solution to selectively radiate or absorb electromagnetic waves. Manifestation of this capability is recorded and serves as a parameter of the matter's nature and content.

These analytical methods differ depending on the area of electromagnetic wave spectrum, in which they operate. In the most long–wave area, beyond the red boundary of the optical light (>700 nm) ***IR spectrometry*** is used. Within the optical and UV ranges the matter's composition is analyzed by polychromatic ***photocolorimetric*** or monochromatic ***spectrophotometry*** methods (***optic spectrometry***). Depending on the source of radiation, absorption, emission and luminescent spectrometry are distinguished. Absorption spectrometry uses an outside source of electromagnetic waves and records the absorption spectra of the analyte. Emission spectrometry analyses the light spectrum radiated by the analyte itself. At last, ***luminescent*** (***fluorescent***) spectrometry analyzes the intensity and nature of short–lasting energized luminescence.

Depending on the kind of action on the analyte, the spectrometry is subdivided into molecular and atomic. Molecular spectrometry is based on absorption by the analyte of electromagnetic waves in the optical spectrum, more rarely IR and UV, and also on the electromagnetic waves–caused

Figure 2.18 Photocolorimeter KFK-3KM intended for the determination of matter concentration in solutions. Provides for the determination of the content in solutions: phosphates, Cl, Cu, Fe, Ag, etc. Wave lengths spectral range is 325–1,000 nm, range of optical density measurement 0–3.

luminescence in the absence of physical action on the state of the solution. This low-temperature spectrometry is intended for molecular or component analysis. Atomic spectrometry is based on the absorption or emission of electromagnetic waves in the optical, IR and UV spectrum of the solution converted in atomic state in conditions of high temperature. High-temperature atomic spectrometry is intended for elemental analysis of water composition.

Molecular Spectroscopy Analysis Methods In hydrochemistry such methods of molecular spectrometry as photocolorimetry, spectrophotometry, fluorimetry, more rarely IR spectrometry and combination dispersion spectrometry are more often used.

a. Photocolorimetric methods are based on absorption of polychromatic light in the optical and near UV spectrum area (Figure 2.18). At irradiation with the light of this spectral range only valence electrons in ions of metals or organic compounds are energized and also electrons carrying charges in complex compounds. These methods are the simplest in molecular spectroscopy and determine the concentration of the matter from the intensity of a solution's coloring. In the case of ***colorimetry*** the color and its optical density (the measure of light absorption by transparent or light reflection by nontransparent objects) are measured visually, and in ***photometry*** they are measured with the help of devices (colorimeters, electric photometers and photocolorimeters).

For this purpose the analyzed component is converted in the form with natural color. Copper is converted into a blue colored compound, Fe^{3+} into red color, Mn^{2+} into raspberry–violet, etc. From the measured value of optical density and calibration diagram, the concentration of analyzed component is then found.

Photometric methods are commonly used for analysis of NH_4^+ content, metals (Al, As, Bi, Cd, Co, Cr, Cu, Fe, Pb, Mn, Mo, Ni, Se, Sn, V, Zn, etc.), and anions (F^-, S^{2-}, SCN^-, CN^-, NO_2, NO_3^-, silicon, phosphorus and organic compounds). Their detection limit at that is usually around 1 μg/l. Besides, these methods determine the contents of such organic compounds as aniline, volatile phenols, methanol, surfactants, polyacrylamide, carbon sulfide, formaldehyde, etc.

b. Spectrophotometry is the measurement of the monochromatic light absorption spectrum in the UV and optical range. The light sources in the optical area usually are tungsten or halogen lamps and in the UV range – deuterium lamps. Monochromatizers, i.e., optical systems letting through electromagnetic wave of the assigned wavelength and providing for maximum absorption effect, are placed either between the light source and analyzed solution or between the solution and signal receiver. Analytical signal is the optical density from which the analyte concentration is determined. For the determination of "colorless" ions, they are colored with appropriate chemicals. For this purpose both inorganic and organic chemicals are used. For recording analytical signals ***spectrophotometers*** are used. Spectrophotometry is used mostly for the determination of medium and low concentrations of organic and inorganic compounds. In particular, in the UV area aromatic hydrocarbons (benzene) are determined, and in the optical, dyes or colored inorganic ions (MnO_4^- has violet color, $Cr_2O_7^{2-}$ – orange) are determined. Selective capability of this technique is not great but may be increased with additional techniques of dividing (for instance, extraction) masking chemicals or simply selecting optimal pH values. The technique's sensitivity depends on the absorbing capability of the component. For analysis of a multi–component system where absorption spectra overlap, special mathematic calculations are used, which increase the analysis accuracy.

c. Fluorimetry is based on the phenomenon of luminescence, i.e., residual glow of the matter after its energizing by outside energy sources. Fluorescence (~10^{-6}–10^{-9} s) and phosphorescence (~10^{-2}–10^{-3} s) are distinguished by the duration of this glow. This glow is associated with the return of electrons from excited state to normal. It does not use heat energy and is often called cold light. Luminescent analysis covers a broad choice of component content determination techniques from simple ions and molecules to high–molecular compounds and entire biologic objects. Either direct luminescence of an analyte or of its derivative is detected. In analytics, more often used is ***photoluminescence***, i.e., fluorescence due to the irradiation in UV and optical spectrum area or *chemiluminescence* – the fluorescence due to chemical reactions. In hydrochemistry matter concentrations are determined from the intensity of fluorescence after irradiation of the solution by UV rays. Devices for this analysis are called ***fluorimeters***. An example is a spectrofluorimeter "Fluorat–02," which supports photoluminescent and chemiluminescent measurements. Wide use of fluorimetry is predetermined by high sensitivity, sometimes at broad range of determined concentrations. Its detection limit is about 10^{-1} μg·l^{-1}. In ground waters such sensitivity enables the determination of Be, Ga, In, Se, U. In complex samples the luminescent detection is sometimes combined with chemical dividing (chromatography, electrophoresis). A drawback of luminescent techniques is the poor selectivity associated with their incapacity to detect analytes in stable complex compounds. Besides, some components in ground water (for instance, at high iron content) are capable of killing fluorescence.

d. IR spectrometry is the recording of light absorption in IR range (>700 nm, i.e., beyond the red boundary of optical light). Sources of radiation are incandescent solid bodies (Nernst lamp or silicon carbide). Monochromatizers enable the identification of part of the spectrum providing the maximum absorption effect. Energy of IR irradiation is only sufficient to excite the oscillatory motion of molecules proper. Therefore, different molecules containing the same bond or the same functional group may yield equal absorption bands. Classic options of this technique do not have high sensitivity and accuracy and provide for mostly qualitative analysis. However, the application of modern techniques for recording IR spectra, especially Fourier's converter, allows substantial improvement of the metrological parameters of this technique. IR spectrometry is applied first of all for identification and determination of organic matter and contents of its structural fragments. Currently IR spectra of more than 20,000 compounds are studied and presented in the form of atlases and tables. IR spectrometry is used for the determination of the content of petroleum products, fats and non–ionogenic surfactants.

e. Collisional activation spectrometry is based on measuring the intensity of dispersed light after its transit through optically nonuniform solution. Most suspended particles and dissolved molecules do not affect the wavelength of dispersed light. But minuscule part of molecules, dispersing light, changes its wavelength. Such dispersion is called ***combination*** (*collisional activation*) or ***Raman***. Analysis of the obtained spectrum allows determining the frequency of natural oscillations of the dissolved molecules and that, in its turn, gives information about their chemical composition and structure. For measurement of this dispersion, monochromatic light is used, as a rule, with a laser ray in the optical range. The combination dispersion (collisional activation) is observed at a right angle to incident light.

Spectrometric and combination dispersion (collisional activation) analyses amend one another. Spectrometry of combination dispersion has some advantage in studies of water solutions. It is also used for molecule identification, determination of individual chemical bonds and functional groups in the molecules and also for the discovery of micro–admixtures polluting the environment. Currently spectrometers are manufactured, which record collision activation spectra of colorless and colored samples whose size is 10^{-4} g (or ml).

Atom–spectrometry Analysis Methods Atom–spectrometry methods are associated with the conversion of an analyte into plasma state in conditions of high temperature. For this a special device is used called an ***atomizer*** to convert analytes in atomic state. These methods to a great extent depend on the atomizer design and plasma temperature. For the determination of ground water elemental composition, researchers usually use atomic absorption, atomic emission and X ray–fluorescent analyses. These

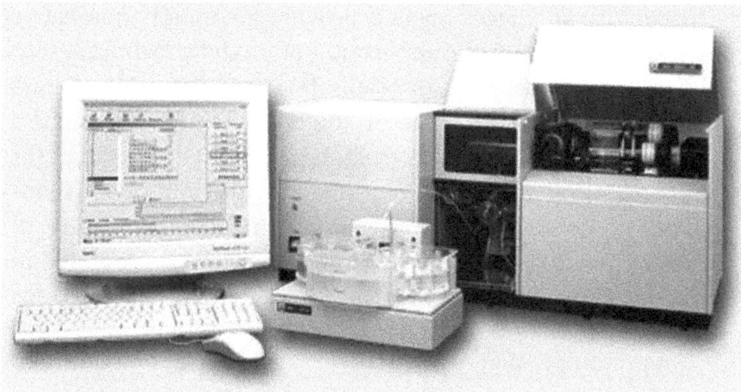

Figure 2.19 Atomic-absorption spectrometer QUANT-2A intended for quantitative elemental analysis of metals (up to 70 elements) in solutions of their salts in natural and waste waters.

analyses do not require plentiful analyzed matter (just 10–30 mg) and complex sample preparation.

a. *Atomic absorption spectroscopy* is based on absorption of the radiation from optical range by neutral free atoms (Figure 2.19). For this purpose the layer of atomic vapor from the sample obtained with an atomizer is subjected to radiation in the rangee 190–850 nm. This radiation also includes the spectrum of the analyzed element. Analyte atoms, transiting to the excited state, absorb light quantums. This absorption is manifested by the appearance of the so-called resonance lines typical of the given element. Change in optical density after running through the atomizer is the measure of its concentration. Depending on the atomizer type these methods are subdivided into two main groups: flame AAS and electrothermal AAS. In the former case for atomization is used the flame from reaction of a combustible gas with oxidants. Solution samples are scattered as small drops in the flame at the temperature of 900–3,000°C. In the latter case solution samples (about 10 μl) are placed in a special induction heater (graphite oven - GO) where they are sequentially dried up, reduced to ash and atomized at temperature 2,000–3,000° C.

   Electrothermal technique is among the most precise and high-performance physicochemical techniques for analyzing liquid samples of various origin. Required sample volume is only 10–200 μl. Currently, this technique can determine the content of 76 elements, mostly metals. Content of nonmetals as a rule is not determined directly. This technique is distinct in high selectivity, sensitivity, accuracy and rapidity of measurements, which makes it commonly usable for analysis of the metal content in ground water composition. Detection limits of metals at electrothermal atomization are about 10^{-5} $\mu g \cdot ml^{-1}$ (Ca, Cs, Ag, Mn, etc.) or 10^{-6} $\mu g \cdot ml^{-1}$ (Be, Zn, Mg, Si, etc.) and are listed in Table 2.11. At flame ionizing for most elements the detection level is better by 2–3 orders of the magnitude (Ag, Cd, Zn, Cu, Fe, Ni, etc.) or higher (Al, Pb, Ba, Ti, V, etc.). Relative error of these techniques, depending on conditions, ranges between 3 and 10%, on average approximately 5%. But AAS is not suitable for the simultaneous determination of several elements in one sample, as each element requires its separate source of radiation. Sometimes for this reason a revolving drum with several lamps is used.

b. Atomic emission spectroscopy (AES) is based on studying the radiation by most energized atoms in a sample. Intensity of this radiation at wave lengths 200 to 1,000 nm serves as the analytical signal. Quality of the analysis is strongly dependent on the atomization technique and the extent of atom excitement. As an atomizer for ground water analysis, most commonly used is either the flame with temperature 1,500–3,000°C or the induction associated plasma with temperature 6,000–10,000 °C.

For the flame-emission photometry, flame of the lighting gas mixed with the air is used. It provides temperature of 1,700–1,900°C and allows the ability to determine about 15 elements (Table 2.11). As sensors for the flame, photometry spectrophotometers are used with automatic recording of spectra and delivery of the results. The flame-emission photometry is the major technique for the determination of alkali metals (Na, K, Li) content.

Currently, however, atomizers with induction-associated plasma are mostly used (Figure 2.20), energized by variable magnetic field from an induction coil. This atomizer yields high temperature (over 6,000°C), which allows almost completely to atomize any analyzed matter and also substantially to lower the detection limits for the elements at sufficiently

Table 2.11 Comparison of element detection levels ($ng \cdot ml^{-1}$) by various methods of atomic spectroscopy.

Element	Symbol	Atomic number	AAS (flame)	AAS (electrotherm.)	AES (flame)	AES (IAP)	AFS GO & LIFS
Aluminum	Al	13	20	0.01	3	0.2	0.6
Antimony	Sb	51	0.1	0.08	200	10	0.1
Arsenic	As	33	0.02	0.08	2000	2	0.1
Barium	Ba	56	8	0.04	1	0.01	2
Beryllium	Be	4	1	0.003	100	0.003	1
Bismuth	Bi	83	0.02	0.1	1000	10	2
Boron	B	5	700	15	50	0.1	2000
Cadmium	Cd	48	0.5	0.0002	300	0.07	0.001
Calcium	Ca	20	0.5	0.01	0.1	0.001	0.08
Carbon	C	6				44	
Cerium	Ce	58			150	0.4	500
Cesium	Cs	55	8	0.04	0.02		
Chromium	Cr	24	2	0.004	1	0.08	1
Cobalt	Co	27	2	0.008	5	0.1	2
Copper	Cu	29	1	0.005	3	0.04	0.03
Dysprosium	Dy	66	50		20	4	300
Erbium	Er	68	40	0.3	20	1	500
Europium	Eu	63	20	0.5	0.2	0.06	20
Gadolinium	Gd	64	1000	8	120	0.4	800

Gallium	Ga	31	50	0.01	5	0.6	0.9
Germanium	Ge	32	50	0.1	400	0.5	100
Gold	Au	79	6	0.01	500	0.9	5
Hafnium	Hf	72	2000			10	
Holmium	Ho	67	40	0.7	10	3	100
Indium	In	49	20	0.02	1	0.4	0.2
Iridium	Ir	77	500	0.5	400	30	
Iron	Fe	26	3	0.01	10	0.09	3
Lanthanum	La	57	2000	0.5	5	0.1	
Lead	Pb	82	10	0.007	0.2	1	10
Lithium	Li	3	0.3	0.01	0.001	0.02	0.4
Lutetium	Lu	71	700		200	0.1	
Magnesium	Mg	12	0.1	0.0002	1	0.003	0.1
Manganese	Mn	25	0.8	0.0005	1	0.01	0.4
Mercury	Hg	80	0.001	0.2	150	1	0.003
Molybdenum	Mo	42	10	0.02	10	0.2	12
Neodymum	Nd	60	600		200	0.3	2000
Nickel	Ni	28	2	0.05	10	0.2	2
Niobium	Nb	41	1000		60	0.2	1000
Osmium	Os	76	80	2	2000	0.4	
Palladium	Pd	46	10	0.05	40	2	40

(Continued)

Element	Symbol	Atomic number	AAS (flame)	AAS (electrotherm.)	AES (flame)	AES (IAP)	AFS GO & LIFS
Potassium	K	19	1	0.004	0.01	30	0.8
Praseodymium	Pr	59	2000		500	10	1000
Rhenium	Re	75	200	10	200	6	
Rhodium	Rh	45	2	0.1	10	30	100
Rubidium	Rb	37	0.3		0.02		
Ruthenium	Ru	44	70		300	30	500
Samarium	Sm	62	500		50	1	100
Scandium	Sc	21	20	6	10	0.4	10
Selenium	Se	34	0.02	0.05		1	0.06
Silicon	Si	14	20	0.005		2	300
Silver	Ag	47	0.9	0.001	2	0.2	0.1
Sodium	Na	11	0.2	0.004	0.01	0.1	0.1
Strontium	Sr	38	2	0.01	0.1	0.002	0.3
Sulfur	S	16	20	10	1600	30	
Tantalum	Ta	73	9			5	
Tellurium	Te	52	0.002	0.03	600	15	0.08
Terbium	Tb	65	600		200	0.1	500
Thallium	Tl	81	9	0.01	2	40	4
Thulium	Tm	69	10		4	0.2	100

Element	Symbol	Atomic number	AAS (flame)	AAS (electrotherm.)	AES (flame)	AES (IAP)	AFS GO & LIFS
Titanium	Ti	22	10	0.3	30	0.03	2
Tungsten	W	74	500		200	0.8	2000
Uranium	U	92		30	100	1.5	
Vanadium	V	23	20	0.1	7	0.06	30
Ytterbium	Yb	70	5	0.1	0.2	0.02	10
Yttrium	Y	39	50	10	40	0.04	500
Zink	Zn	30	0.8	0.0006	1000	0.1	0.0003
Zirconium	Zr	40	350		1000	0.06	
Платина	Pt	78	40	0.2	2000	0.9	300
Sample volume, ml			5–10	0.01–0.1	5–10	1–10	0.01–0.1
Max salinity $g \cdot l^{-1}$			30	200	30	10–100	10
Detection limit			$1–10^3$	10^{-2}–0.1	1–100	0.1–10	
Detection limit, ng/ml			0.01–2000	0.0002–30	0.001–2000	0.001–44	0.0003–2000

Note. AAS is atomic–absorption spectroscopy, AES is atomic–emission spectroscopy, IAP is induction–associated plasma, and AFS is atomic–fluorescent spectroscopy. GO is graphite oven, and LIFS is laser-induced fluorescence spectrometry.

Figure 2.20 Atom-emission spectrometer with induction - associated plasma by "SPECTRO ANALYTICAL INSTRUMENTS" (Germany). Intended for quantitative analysis of a great number of elements (metals, sulfur, phosphorus, chlorine, etc.). Simultaneous determination of several elements in the solution is possible.

high values of correctness and reproducibility. The main advantage of this technique is that it allows simultaneously to determine the content of a whole number of metals (Ag, Al, As, B, Be, Br, Ca, Cd, Co, Cr, Cu, Fe, K, Mg, Mo, Mn, Na, Ni, Pb, Sb, Se, Si, Sn, Sr, Tl, Zn, V). For this reason the AES with induction-associated plasma becomes typical multi-element analysis technique. Its advantages are accuracy, speed and high sensitivity (for alkali elements 0.01 $\mu g \cdot ml^{-1}$, for alkali-earth 0.1 $\mu g \cdot ml^{-1}$). The operating range of concentrations is 1 to 1,000 $\mu g \cdot ml^{-1}$ (Table 2.11).

c. Atomic fluorescence spectroscopy (AFS) is based on measurements of atom fluorescence intensity after their irradiation. For this the sample in an atomizer is put in the state of atomic vapor and its fluorescence is forced by an outside source of light. A spectrometer analyzes the fluorescence spectrum whose intensity reflects the content of elements. Variability of this technique is determined by the type of atomizer and of the light source. Mostly flame or electrothermal (graphite ovens) atomizers are currently used. The sources of radiation are high-intensity discharge lamps (commonly with hollow cathode) or variable frequency lasers (laser-induced fluorescence spectrometry - LIFS). Radiation intensity of excited atoms in the optical range serves as the analytical signal. Mostly X-ray-fluorescent

Figure 2.21 X-ray-fluorescent spectrometer **"SPECTROSCAN MAX-GV"** for the determination of chemical elements content in various matters including in dissolved state. Detection limits: Na – $1 \cdot 10^{-1}$ %, Mg, – $1 \cdot 10^{-2}$ %, Al – $1 \cdot 10^{-3}$ %, Si – $5 \cdot 10^{-4}$ %, P – $5 \cdot 10^{-4}$ %, Cd, Pb – $5 \cdot 10^{-4}$ %, S, Ti, V, Cr – $1 \cdot 10^{-4}$ %, Co, Ni – $5 \cdot 10^{-5}$ %.

spectrometers are used in recent times (Figure 2.21). For some elements (Zn, Cd, Hg, etc.) relative detection limits reach ~10^{-5}–10^{6} %.

2.2.4.3.5 Mass-spectrometry Methods

Mass-spectrometry is based on dividing ionized atoms or molecules of the analyzed matter by the mass. For this, the matter is ionized, then the ions are divided in magnetic field by the value of their mass to charge ratio, and the derived distribution of their amount relative to mass is recorded as ***mass-spectrograms***. For dividing ions, magnetic or quadrupolar dividing are most commonly used. In the former case ion dividing occurs in a uniform magnetic field, in the latter one, in a high-frequency electric field created between four parallel bars - condensers. Depending on the ionization method and energy, ***analytical*** and ***elemental*** mass-spectrometry is distinguished.

"Analytical" is called mass-spectrometry with incomplete decomposition of compounds into atoms and intended for qualitative or quantitative analysis of components. Under the mildest ionization (7–13 electron-volt) compounds almost do not decompose. They lose one electron and form a positively charged cation-radical (molecular ion). Under such ionizing molecular analysis is achieved, at which the molecules are distinguished mostly by their mass.

With increasing ionization energy to 70 electron-volt the disintegration of cation-radicals occurs, which is their fragmentation into smaller parts. Mass-spectrograms of these ionized fragments are the basis for molecular quantitative analysis of inorganic gases. Organic compounds under such ionization decompose into charged fragments whose relative contents are determined from the mass-spectrogram. Relative content of fragments define the nature and amount of organic analyte.

"Analytical" mass-spectrometry has great significance for the identification of complex organic compounds. Analysis of such compounds is hindered by the complexity of obtained mass-spectrograms. For improving the selectivity, first, ***tandem mass spectrometers*** are used, which include two mass-analyzers with different ionization energy. The first one differentiates molecular ions after mild ionization, the second one studies derived fragments after additional harder ionization. Second, ever more often the analytical mass-spectrometry is combined with different kinds of chromatography and capillary electrophoresis. Such combination of different analysis techniques is called ***chromato-mass-spectrometry***. Most common is the combination of mass-spectrometry with gas-liquid chromatography where helium is the carrier gas. This technique is suitable only for volatile components.

Mass-spectrometry is widely applied for molecular analysis of water solutions. Its role is especially great in the determination of the dissolved gas and organic components content. For this is used mild ionization usually with electron impact. Most commonly mass-spectrometry is used for the identification and analysis of the content of complex organic compounds. The analytical mass-spectrometry in combination with chromatography provides for high sensitivity, selectivity and universality in analyzing a wide circle of gas and organic compounds up to protein molecules.

The elemental mass-spectrometry deals with atoms and conducts elemental or even isotope analysis of the matter. Charged atoms are produced in atomizers (by spark discharge, in induction-associated plasma, by a laser, etc.) The spark discharge is used for this purpose ever more often. It provides quite high selectivity and sensitivity (detection limit up to 10^{-12} g), and allows the determination of simultaneously up to 60–70 elements. The isotope mass-spectrometry has especially high sensitivity.

Mass-spectrometry is the major technique for the determination of isotope composition for the individual elements in water composition. For analysis of light elements' isotope composition (C, H, O, S, N, etc.) the electron impact ionization is used. For the isotope analysis of heavy elements, ionization is applied in induction-associated plasma. Magnetic mass-analyzers are used practically in all kinds of isotope mass-spectrometers. Isotope analysis, as a regular rule, requires no less complex preliminary operation of separating of analyzed element or its compounds in pure form from the studied natural medium.

2.2.4.4 Selection of a Laboratory and Executants

Many parameters of water, natural gas or organic matter composition may be determined only by instrumental methods in stationary laboratory conditions by qualified experts. The selection of these experts and their analytical techniques, i.e., analytical laboratory, should be made before field operations, which would allow determining in advance, and guarantees analysis quality, minimum volume, size, and nature of the sample preparation. This would also facilitate correct selection of vessels for the samples by their quality, volume and number. As a rule, samples for special analyses are collected separately in different vessels.

In selecting an analytical laboratory, it is necessary to take into account such parameters as competency of its analysts, range of determined concentrations, technical accessibility, price and quality of measurements of the needed parameters.

Laboratory competence assumes favorable conditions, appropriate equipment and experienced personnel for assuring the required quality of chemical analyses. The official indication of this competence is by accreditation diploma and certificates of quality. ***An accreditation diploma*** is an official document confirming that the laboratory got through accreditation, i.e., official maximum objective and qualified evaluation of its competence. Technical competence means the capacity of the laboratory to perform the stated types of chemical analysis with required accuracy. For the recognition of technical competence is necessary the availability in the laboratory of the equipment (and free access to it), chemicals and documents for the conduct of quantitative chemical analyses, a system of analysis quality control, and qualified personnel. In the process of the accreditation, the laboratory is checked for independence of any external pressure in order to eliminate any influence on the produced results.

For this reason in order to be acknowledged as an independent laboratory it is usually required for it to have legal status and the absence of liabilities, which might cause doubts in its objectivity. The accreditation diploma is accompanied with a list, which determines the accreditation domain. This list usually contains the objects of analysis (water, rocks, etc.), components whose content is measured, and range of their concentrations. The accreditation indicates that analytical methods and their results have systematic objective quality evaluation and correspond with requirements of the official technical standard documentation.

For laboratories conducting routine studies, accreditation is usually required as certification of individual analytical techniques, which is confirmed by special documents (***quality certificates***). The certification is conducted voluntarily but allows the laboratory to form and strengthen its positive image and provides advantages in getting state orders and tenders, and substantially simplifies the process of obtaining licenses or permits.

Data of the ground water composition play ever greater role in solving issues of ecological safety. In connection with these ever more often are required legal guarantees of quality of provided analysis results. Such guarantees may be given only by laboratories possessing accreditation diplomas and quality certificates. In particular, according to SanPiN 2.1.4.1074–01, for the conduct of a laboratory study (measurements) of drinking water qualities, only the laboratories are allowed which are metrologically certified and approved by Russia's Gosstandard or Russia's Ministry of Health Care.

Measurement of analyte concentration range describes values, which may be determined in laboratory under the indicated technique. Obviously, this range must cover the interval of analyte concentrations, which is expected in studied ground waters. The measurement range has two boundaries. As a rule, the main attention is devoted to the lower boundary, which is associated with detection limits and measurement, and to a smaller extent to the upper one. If natural analyte concentrations turn out to be lower than the range of measured concentrations, analyte concentrating or extraction is applied. If natural concentrations turn out to be higher than the range of measured concentrations, the samples are diluted with distilled water. In either case appear an extra operation and the associated error. The order of individual elements' content values in low salinity ground water is listed in Table 2.12. It includes logarithms of mean concentrations (nanograms per liter) for individual elements in weathering zone waters in leaching and salinization provinces. The range of measured concentrations according to this Table is determined as the interval between values 10 times less than and 10 times more than values in the Table. The lanthanides are usually contained in concentrations less than 1 micro–gram per liter.

Technical accessibility here means the possibility for a client to satisfy requirements of the analytical laboratory, namely, to provide for analyses the samples of necessary volume and quality. As a rule, for reliable analysis, not the water volume, but the total amount of the analyte is important. It is necessary to have greater than 0.1 g for macro–analyses, less than 0.01 g for micro-analyses, and less than 0.001 g for ultra-micro-analysis of the analyte. At very low concentrations it is sometimes necessary to concentrate the analyzed matter. For a qualitative analysis the laboratory may request either too large volume of samples (up to tens of liters) or a preparation inaccessible for technical reasons.

Chemical analysis price in the selection of analytical laboratory is sometimes very important. The less the financial means of the client, the higher

Table 2.12 Logarithms of mean concentrations of chemical elements in low salinity ground water (of leaching – salinization provinces) (Shvartsev S.L., 1998).

Element	Concentration logarithms, $ng \cdot l^{-1}$	Element	Concentration logarithms, $ng \cdot l^{-1}$	Element	Concentration logarithms, $ng \cdot l^{-1}$
Ag	2.4–2.6	F	5.3–6.2	Ru	<3
Al	5.2–5.6	Hg	1.6	S	6.1–8.0
As	2.9–3.3	I	3.3–4.0	Sb	3.0–2.9
Au	0.3–1.1	In	<3	Sc	2.0
B	4.3–5.4	K	5.9–7.3	Se	2.1–3.3
Ba	4.0–4.5	La	3.0	Si	6.6–7.2
Be	1.7–2.5	Li	3.3–4.6	Sn	2.5–2.7
Bi	<3	Mg	6.7–7.7	Sr	4.3–5.7
Br	3.9–5.4	Mn	4.1–5.1	Ta	3
Ca	7.2–7.9	Mo	2.9–3.6	Te	3
Cd	2.5–2.6	Na	6.8–8.4	Th	1.8–2.9
Co	2.5–2.8	Nb	2.7	Ti	3.5–4.8
Cr	3.4–3.6	Ni	3.3–3.7	Tl	3
Cs	2.7–2.8	P	4.4–4.8	U	2.4–3.6
Cu	3.4–4.1	Pb	3.2–3.8	V	2.8–3.4
Fe	5.4–5.9	Pd	<3	W	4
Ga	2.4–2.8	Pt	<3	Y	4
Ge	3	Ra	2.9–3.5	Zn	4.4–4.9
Hf	<3	Rb	3.2–3.3	Zr	3.1–3.1
Cl	6.7–8.4	Re	3		

the significance of this factor. As a regular rule, the price of an individual analysis depends on the determined parameter, the measurement accuracy and the time expended for the analysis. The cheapest operations are the determination of individual organoleptic properties and pH. If their price is a unit, the measurement of CO_2 and O_2 contents, hardness, permanganate oxidability and general microbial amount will be 2 units of the concentrations of mineral components, and TOC and coli-index will be 5–10 units. Substantially higher is the price of organic compounds' analysis, especially halogen-containing (25 units) and pesticides (34 units). The most expensive operations are the determination of the number of Lamblia cysts (150–230 units) and identification of individual organic compounds by the gas chromatography–mass–spectrometry (367 units). Thus, the determination in 2009 by ZAO "ROSA" of minimum list of parameters necessary under SanPiN 2.1.4.1175-02 for the evaluation of general water pollution in wells and shaft wells and recommended for annual prophylactic control of water quality was around 4.0 thousand rubles, whereas the primary study of water qualities under the expanded list was 17 thousand rubles.

If the laboratory services are acceptable by legal, technical and financial parameters, the client should become familiar with the quality of accomplished measurements, namely, with the selectivity, sensitivity and quality parameters of the analyses results.

Selectivity of chemical analysis describes the effect of outside components in the solution on the result of measuring concentrations of an individual analyte. This is especially important relative to underground waters, which are complex multicomponent solutions. First, in its composition the analyte may be part of different more complex and more stable compounds, which are not taken into account by the analysis. Before the analysis the analyst tries to destroy all such compounds and bring the analyzed component to a single simple form. This is not always possible. Second, components may be present in the ground water similar to the analyte, which, for instance, coprecipitating with it or interacting with its chemical, may thereby overstate its content. Such ions are called ***impeding***. The methods, reactions and chemicals, which in this sample determine the content of only one analyte, are called ***specific***, and which allow measuring the contents of several different analytes at once, ***selective***. High selectivity is achieved by the right selection of chemicals, sequence of conduct and conditions of analyses. At the low selectivity this correlation increases.

The selectivity issues should be solved in advance together with analysts based on assumed water composition in order to search for optimal analysis procedures. On this depends which components are impossible to determine in one sample, how to prepare them in the field, how to safe keep, and in what order to analyze. Components most sensitive to the change of conditions are determined first. For instance, the loss of CO_2 at the opening of samples may cause the formation of carbonate deposit, may affect the Ca^{2+} content and salinity and pH values, specific electric conductivity, tittered hardness and alkalinity. The precipitation, in its turn, may cause the loss of codepositing micro–components. For this reason all samples before analysis must be visually checked for the presence of a deposit. If there is a deposit, the analysis results will not reflect the samples as of the moment of testing. Such samples require special preliminary treatment in laboratory before the analysis. The nature of this treatment depends on the nature of the deposit and of the components whose content need to be determined. After the treatment the analysis may be performed with reasonable accuracy. Besides, it was noticed that due to changes in water's chemical composition, with an increase in duration of analyses the reliability of analytical results declines. All determinations are desirable to do within 1 week after opening the sample. Sometimes for developing the optimal analytical procedure preliminary control analyses are required.

Sensitivity of the analysis technique in any laboratory describes detection limit (DL) and measurement or quantification limit (QL) values of the analyte concentration. Analytes with concentrations below these limits are impossible to detect reliably. In order for a study to be successful it is necessary for the DL of the technique in qualitative analysis and QL in quantitative analysis to be much lower than the analyte concentrations in samples. For this it is necessary to have some idea of anticipated concentration values of the studied analytes in ground waters. In the absence of such data, it may be useful to refer to Table 2.12 listing concentration logarithms (nanograms per liter) of individual elements in waters of weathering zone in leaching and salinization provinces. Sensitivity of the analysis technique should be selected by one order of the magnitude lower of the averages in this Table.

Analysis results' quality parameters in individual laboratory are characterized by the values of accuracy, repeatability (convergence), within-the-laboratory reproducibility and systematic error for each technique.

Technique repeatability in laboratory describes its minimum precision at analysis of the same samples by the same executant under conditions of strict observance of the procedural repeatability. ***Within-the-laboratory reproducibility*** is maximum precision of repeated analyses of an analyte from the same sample in individual laboratory (different time, different analysts, different batches of chemicals of the same type, etc.). These precision parameters define the range of random errors at analysis of samples in a given laboratory. The sample collection and their transport to the laboratory facilitate the magnification of these errors. For this reason a laboratory should be elected, which ensures analysis technique reproducibility at least by one order of the magnitude less than the studied range of analyte concentrations in studied ground waters. A random error may be noticeably lowered by increasing to 2 or 3 the number of replicate sample analyses.

The laboratory systematic error in a particular technique is the difference between mathematic expectation of the analysis results for the same sample, derived in the given laboratory, and sought-for true (or accepted reference) significance of the measured parameter. Laboratories must themselves identify and eliminate systematic errors by way of perfecting techniques and introducing corrections. Systematic errors of laboratories are one of the basic parameters of their incompetency, and not a single laboratory would admit erroneousness of its results. For this reason the hydrogeochemist should himself check the laboratory analyses results' correctness. For that at individual most responsible objects should be collected split samples of water for analysis in 2–3 different laboratories by different methods.

In selecting a laboratory by the accuracy of analysis technique it is necessary to take into account that the testing accuracy in the final analysis almost equally depends on errors of analysis and of field sample collection. Both should be of approximately equal order as out of economic consideration there is no sense in increasing accuracy of measurements until more reliable methods of sample collection are developed, and vice versa.

2.2.5 Field Testing Protocol

Field testing is the most responsible, costly and logistically complex part of the hydrogeochemical study. In the monitoring studies field testing is repeated at **observation** points with planned periodicity. In single-time studies it is numerously repeated at different objects within a rigidly restricted time interval (1–3 months).

Testing of each water-show is a complex operation, which includes a number of procedures performed in the field in uncomfortable, sometimes difficult climatic conditions, and as a rule, by different people. That makes it difficult to observe strict identity of all testing procedures. However, under GOST P 5725-1-2002 – 5725-6-2002 measurement of the same parameters within the same studied object must be performed in a strictly identical way, i.e., measurements methods must be standardized. To comply with this a regulating document is needed, which would determine in detail the nature and procedure of the testing. Such document is the ***sampling protocol*** – the instruction containing detailed description of the content, sequence and recording of the sampling procedure.

Testing protocol is put together in consideration of the task assignment, climatic and hydrogeological environment in order to minimize errors associated with field works. It allows unifying field testing technique and lower influence of changing executants.

Testing protocol may include 5 items, which determine the sequence and content of the performed operations:

1. Descriptions of the testing object conditions and status.
2. Preparation of the object to testing.
3. Extraction of the water on the surface.
4. Field measurements.
5. Collection and preparation of water samples.

Objectives and contents of these testing items are represented in Table 2.13. Each point includes the preparation and performance of the necessary procedure with simultaneous recording of conditions, qualities and results in a ledger. In any testing of ground water, especially of their contamination halo, the testing should begin with wells, located upslope of the ground water level in order to minimize the possibility of cross-contamination of water samples by instruments.

Description of the conditions and state of testing object begins with the recording of the sampled water point, i.e., assigning a number to it. At that the nature of the water show (spring, shaft well, well, etc.) is noted. If the testing point is not on the map it should be tied-in and plotted on the map. It is recommended to use GPS or GLONASS. Description of the water point should include the characteristics of topography and surface water, information about lithology, stratigraphy of enclosing sediments, depths of sampled water occurrence and the design and state of tapping, whether in shaft well or well. Additional information depends on the testing purpose.

Preparation of an object for testing depends on its nature. In case of a surface water spring it is sufficient to clear the points of testing of outside objects and to create conditions comfortable for collection of samples and for field study. If the water occurs at depths, in shaft well or well, it is necessary to prepare and wash the instruments for its extraction (pump, hoses, samplers, etc.).

Table 2.13 Main points of ground water testing protocol.

Protocol items		Content	Recommendations
1	Description of conditions and state of testing object	Geophysical and geologic tie–in, measurement of depths to water and its flow rate, data recording in the ledger	For springs is measured flow rate, for wells and shaft wells – depths to water level with accuracy ± 0.3 cm.
2	Object preparation to testing.	Cleaning up spring mouth, shaft well, removal of stagnant water or flushing the well. Recording flushing conditions in the ledger.	Cleaning up wells by water withdrawal (> 2 volumes of well) until stabilizing (±10%) of controlling parameters (pH, T, Ω^{-1}, Eh) while flushing.
3	Water extraction on the surface.	Lifting water on the surface by pump or sampler. In case of water natural flow on the surface this procedure is eliminated. Used methods are recorded.	In pumping water out the rate of about 0.1 m^3 per day is desirable for minimization of suspension and loss of volatile components.
4	Field measurement	Determination and recording of organoleptic properties and results of water properties and composition field measurements.	Measurement is done directly at points of water coming to the surface.
5	Collection and preparation of water samples.	Collection and preparation of work and control water samples by disruption of the sample chemistry. The samples are recorded in the ledger and supplied with lables.	First collected and prepared are the samples for volatile components without filtration and then the samples requiring filtration.
6	Safekeeping of the sample / Transport	Packing the samples for transport or placing them under required conditions (refrigerator) in order to preserve and minimize changes in their composition.	Sample labels must be doubled. Observe acceptable period of sample safekeeping.

At the collection of ground water from shaft wells or wells it is necessary to take into account their possible contamination. In old abandoned wells and shaft wells contamination may be associated with the presence of stagnant water. It differs from the underground water due to the interaction with materials of wells and outside objects falling from above, with atmospheric oxygen, etc. Moreover, this stagnant water may penetrate the near–hole reservoir area and form a contamination train. In new wells the testing is usually hindered by the contamination caused by the drilling instrument, drilling mud, material of the gravel filter and even by the casing corrosion. This influence may be strengthened by oxidation–reduction processes and ion exchange.

For this reason before collecting water samples from shaft wells and wells they are carefully run for several hours. That allows eliminating outside water and turbidity at well filters. The water is pumped at some optimal regime, which eliminates the inflow of water from other horizons. In most cases it is sufficient to pump out 3–5, sometimes up to 10 well volumes. The best way to control qualities of the cleanup is monitoring easily measured parameters of water qualities such as temperature, electric conductivity (EC), pH or Eh. Control by the pH value is convenient at elevated content of CO_2 in sampled water. In practice cleanliness of wells is considered OK if these controlled parameters are stable for 2 successive volumes of wells. Example in Figure 2.22 shows that stationary conditions are reached after removal of greater than two volumes of wells. For pumping out in the field, it is necessary to have a powerful pump, hose and energy source.

Ground water extraction on the surface is required only at their deep occurrence in shaft wells or wells. When water is on the surface in dig holes,

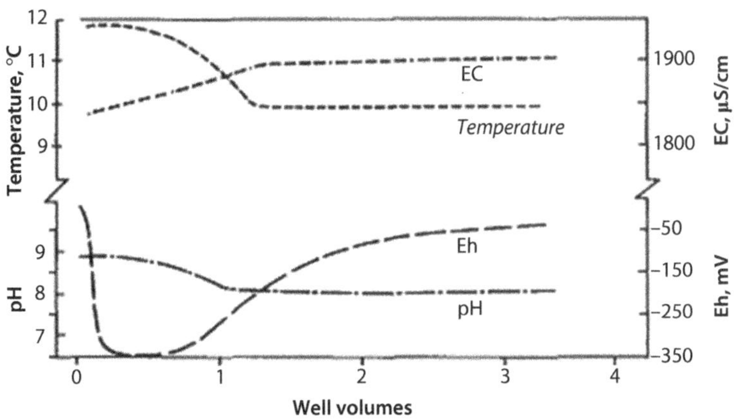

Figure 2.22 Change of underground water chemical properties during well flushing. EC is specific electric conductivity.

in springs or due to natural flow it is sufficient to clean or equip it and make a simple tapping. The deeper the water occurrence depths, the more difficult it is to sample. There are two basic ways to lift the underground water on the surface: by pumping and by sampler. In the former case the water is lifted on the surface by pumping, and all used equipment (the pump, tubes, hoses, etc.) are cleaned up together with the well at the time of its preparation to testing. The amount of water usually is unlimited, and water cleanliness is possible to control by monitoring its properties while flushing. In the action of testing it is recommended to maintain the lowermost withdrawal rate, up to 0.2 $м^3 \cdot day^{-1}$ (100 $ml \cdot min^{-1}$) in order to avoid capturing outside colloid particles. In the latter case a sampler is run to the depths of water collection, which is filled up with water and raised to the surface. Working with samplers, water volume is usually limited, which noticeably increases the risk of its contamination. Besides, at any non–hermetic water extraction it is possible to lose dissolved components due to the decline of temperature and pressure. Lost, first of all, are volatile components. Their loss increases with water gas–saturation increase with depth. The cunning of this situation is that such loss is possible to notice and estimate only using a reliable separator at the wellhead or special hermetic samplers.

Samplers are commonly used for the collection of different ground water samples. Thus, the danger occurs of cross–contamination, i.e., mixing of different waters. Equipment–associated outside pollutants may be used in laboratory chemicals (phthalates, methylene–chloride, acetone, etc.), oily materials, and sometimes chlorine–containing organics. For this reason the used samplers and their instruments should be carefully washed before testing with fresh water and rinsed with water not containing the analytes. The best for rinsing is distilled nonionized water. But it is also possible to use other water where the absence of the analyte is confirmed by chemical analysis. Rinsing is performed 2 – 3 times. In the last rinsing the sampler is filled up with distilled water, which then is treated as the collection of work samples with the analyte, i.e., filtered, prepared and safe–kept until analysis. These samples of the last rinse are ***instrumental blank samples*** (***rinse or cleaning blank***). They are prepared in the field at the time of testing, and in the documentation and safekeeping conditions they must not differ from work samples of the same analyte. They are necessary to control any contamination from the used equipment, especially for the determination of organic pollutants. Comparison of blank instrumental sample analysis results and blank field samples allows exposing errors associated with a contaminated sampler. Instrumental blank samples are collected for each sampler separately in the number of at least 1 per 20 testings. However, it is better to have at least 3 blank samples per each sampler. The number of instrumental blank samples may be smaller if the instruments are for single–time use.

Field measurements are recommended to be conducted prior to collection of samples directly at water-point, immediately after a careful preparation of the source or lifting water from shaft wells or wells. For this it is necessary to have at hand all chemicals and devices, preliminarily calibrated. Measured in the field are usually the parameters, which are either unstable (temperature, gas–saturation, pH, Eh, content Fe^{+2}, F^-, NO^{3-}, O_2, CO_2, etc) or necessary for updating the testing and analysis techniques or are used for control of laboratory sample analyses (electric conductivity, Cl^-, SO_4^{2-}, general hardness, etc.). Values of pH and Eh are needed for evaluating the existing form of micro–components with variable charge and techniques of their conservation. Specific electric conductivity gives the idea of water salinity, which is important for the determination of optimum volume of samples for micro–components and helps to control the errors of general water analysis. Methods of field analysis and their acceptable errors are regulated by GOST 24902-81 "Water domestic/potable. General requirements to field analysis methods". Sometimes field works include sanitary–micro–biologic analysis of potable water. Such need occurs when the delivery of samples to a stationary laboratory within 6 hours is impossible. These field analysis methods are regulated by GOST 24849–81 "Water potable. Field methods of sanitary–microbiologic analysis". Under this, GOST analyses are performed either with portable laboratories on the spot or in mobile laboratories (in trucks, ships, etc.).

In field measurements errors associated with the sample collection and their safekeeping are eliminated. Errors of field measurements are associated mostly with used measurement methods. These errors may be estimated and controlled on the spot right in the field. Random error is determined from several repeated measurings (5–20) of a specific parameter in the same water. Systematic error is estimated either relative to calibrating solutions or relative to the analysis results of the same parameters in stationary laboratories. Field detection limits and analyte concentration measurement in field conditions are assumed equal to the similar laboratory parameters.

Collection and preparation of water samples are the most responsible testing procedures. Sampling should be done as close as possible to the points of water exiting on the surface. Often even at one water pointe several different samples are selected simultaneously for different parameters. This is associated not only with the need of analysis in different laboratories (chemical, gas, micro–biologic, isotope, etc.). Some dissolved components obstruct one another at analysis and require separate testing.

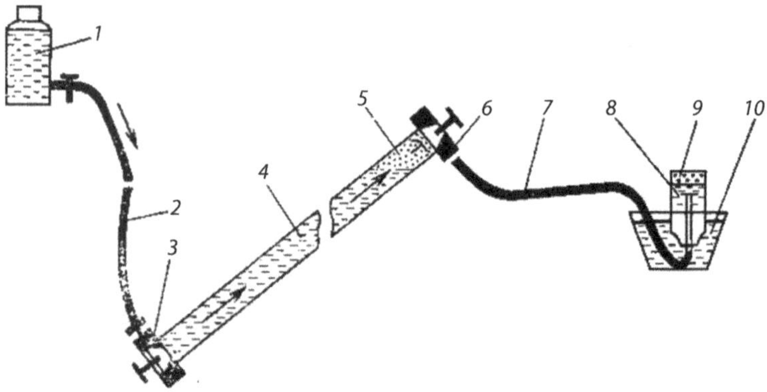

Figure 2.23 Degassing of downhole sampler by "pressing up" method: 1. Bottle filled up with saturated table salt solution; 2. Rubber hose; 3. Lower double-valve adapter; 4. Sampler; 5. Sampler's dead space; 6. Lower adapter; 7. Hose; 8. Metering vessel; 9. Dead space of metering vessel; 10. Basin (V.N. Kortsenshtein, 1991).

Samples are recommended to collect in the following order:

1. Natural gas and volatile organic matter.
2. With extraction of organic matter including petroleum products, organic acids etc..
3. For total content of individual metals.
4. For the content of individual dissolved metals.
5. For microbiological analysis.
6. For analysis of inorganic anions.
7. For total macro component analysis.

As this priority list shows, samples of natural gases and water for most volatile organic components not requiring filtration should be collected first. During the collection of natural gas samples it is very important to eliminate their contamination with air. With the use of downhole samplers this is achieved with special adapters (Figure 2.23). In the second turn are collected samples for nonvolatile organic components. Extraction of these components often requires a substantial volume of water. In the case of collecting samples for micro-biological analysis it is very important to maintain sterility of the vessels, samples and used chemicals. In the USA water samples for such analyses are sometimes collected in containers, carefully blown through with molecular nitrogen, in special hermetic chambers filled up with the same gas. The last collected are samples, requiring preliminary filtration in the field (for instance, metals, alkali and anions). The volume of samples depends on the technique of analysis and may vary between 10–15 ml and 1–15 l. Volume of individual samples for analysis of macro components usually does not exceed 0.5 l.

Samples for the same parameter may be test and control. ***Test samples*** are intended directly for measurement of studied water properties and composition parameters. Each test sample must have 1–2 duplicates. A ***Duplicate test sample*** is a spare test sample. This is necessary not only for insurance for a case of disruption of the integrity or destruction of the sample but also for a possibility to repeat analysis results in case of an unexpected result. Duplicate samples may be used as a repeat control.

Control samples are used for estimating measurement quality not in test samples but with help of test samples. For this purpose are selected samples: repeated, field blank and spiked.

Repeated sample is no different from duplicate samples in the nature of collection but is mandatorily subjected to analysis. It is recommended for each measured parameter for one field season to collect 2–3 repeated samples in more than 10% of all testing points so that their total number is at least 20 samples.

Field blank sample is intended for the control of errors in analyte content in work samples and for the estimate of testing sensitivity as a whole. For this reason field blank samples are prepared in the field using the water without the analyte, safe-kept and analyzed together with test samples. It is recommended to collect one sample a day for each tested parameter so that their total number was at least 3.

Spiked-sample may be blank and test. If the analyte solution is added to a blank sample, the result is ***blank spiked-sample***, and if is added to a test sample, the result is a ***test spiked sample***. It is recommended at the collection of a test spiked-sample to use spiked-solutions with analyte concentration much higher than expected in the tested water. Such spiked solution may serve as a sample for the comparison. It is introduced into the vessel for samples first in such amount that its volume is at least 1 ml and no greater than 1% of the sample volume. In order for the spiked-sample volume at the addition to change negligibly little, the analyte concentration in the spiked-solution must be quite high. Blank spiked-samples may be prepared with different analyte concentration but at least 10 times its laboratory detection limit. The volume of added spiked-solution may be approximately estimated from the following equation:

$$V_{\mathrm{sp}} = V_{\mathrm{pr}} \frac{C_{i,\mathrm{pr}}}{C_{i,\mathrm{sp}}}, \tag{1.70}$$

where $C_{i,sp}$ and $C_{i,pr}$ are analyte concentrations in the spiked-solution and in the spiked-sample, respectively, V_{sp} and V_{pr} are volumes, being mixed, of the spiked-solution and blank or test sample.

Test and their corresponding control samples are prepared, recorded, safe–kept and analyzed equally.

The preparation may involve filtration, conservation and concentrating. In a case of testing for estimating water suitability for drinking, the sample preparation procedure is strictly regulated by GOSTs P 51592-2000 "Water. General requirements to sample collection" and P 51593-2000 "Water potable. Sample collection". These GOSTs include general requirements to the collection of samples for measuring the most important parameters of water qualities. However, in special studies such testing is necessary to perform in consideration of analyte properties and specifics of its analysis. Thus, the water sample collection technique for special study should be coordinated with the recommendations by analysts from the selected laboratory.

For the collection of samples it is necessary to prepare vessels and stoppers resistant to external physical and chemicaly actions and providing for reliable hermeticity, and also chemicals and extractants necessary for dividing or concentrating of the analyzed components. The vessel material must be resistant to long action of water and must not introduce into it outside components. Samples for metals may be collected in a plastic container but stainless steel and rubber are not recommended, and even more so not recommended are corrodible metals. Samples for organic matter should be collected in glass or Teflon vessels. Untransparent or darkened containers protect them against light and are used for samples for pesticides, herbicides and other complex organics. Very convenient are 40 ml phials with screw–up lids with Teflon washers. Often used are lids with Teflon membranes, which are punched through with the syringe needle.

Vessels and stoppers must be preliminarily carefully washed and held in water several days. It would be nice to have this water acidized with nitrogen or hydrochloric acid to pH lower than 2. Such soaking facilitates the decrease in adsorption capability of vessel walls. Sometimes they are washed with acetone, chrome mix (solution of H_2SO_4 and $K_2Cr_2O_7$) or Komarovsky's mix (solution of HCl and H_2O_2) and rinsed in distilled water. Vessels for samples for phosphates, silicon, boron and surfactants should not be cleaned by detergents.

The preparation usually begins with the removal of coarse suspension or with filtration for the removal of suspended matter, deposit, alga and microorganisms. A paper filter is used for filtration. Membrane filter should be used carefully as it may absorb or introduce in the sample heavy metals or organic matter. If preliminary filtration is required for several samples at once, it is conducted before decanting water in vessels. This allows saving time, prevents mass transfer between water and the deposit in the sample and provides for identity of the samples for a subsequent analysis .

Obviously, the most reliable results provide analyses performed right at the moment of testing or 4 hours after the collection. If the transport of samples in a remote laboratory is needed, they are conserved. The conservation nature depends on the purpose of the sample, type of analyte and duration of the sample's safekeeping (Attachment I).

For the conservation of many analytes is sufficient to acidify the samples to pH 2–3 or alkalize to 11. Addition of acids facilitates slowing down micro–biologic processes, preservation of metal oxidizing degree and also minimizes the precipitation of hydroxides and adsorption on the vessel walls. For suppression of microorganisms' activity to the samples for components, participating in metabolism, are added antiseptics (chloroform, thymol, etc.). In some cases the samples are added with special chemicals in order to bond the analyte, making its conent stable.

In many cases the vessel is filled up with samples to the very stopper, not leaving the air. This restricts the interaction between water and O_2 of the air and prevents sample shaking in transport. However, in the samples for microbiologic studies it is recommended to leave air under the stopper. This restricts a possibility of sample contamination and allows shaking and stirring the sample before analysis. Not filled up to the stopper are also samples which are conserved by freezing.

At last, the plugged samples in most cases are cooled down to 2–4°C and held on ice or in a refrigerator. More rarely they are frozen at a temperature below –20°C in a freezer.

Conservation in most cases allows safekeeping the samples for several days. For longer safekeeping of samples (a month or longer) freezing or concentration is used on analytes by evaporation, degassing, extraction, distillation, etc.

In any case testing and long safekeeping of samples must not affect the analysis results. If there are several options of sample preparation and safekeeping, they should be tested preliminarily and compared with the help of blank samples. In preparation of samples, especially with the use of dangerous devices or chemicals, safety procedures should be observed.

The sample collection process is recorded in the ledger or act, which, under GOST P 51592, must contain at least the following data:

1. location with coordinates and name of the water point;
2. collection date;
3. collection technique;
4. collection time;

5. weather conditions at the sample's collection;
6. water temperature at sample's collection (if needed);
7. technique of the preparation for safekeeping (if needed);
8. purpose of water testing;
9. other data depending on the purpose of sample collection;
10. position, name and signature of the responsible executant.

Each sample must be accompanied with a label. It should indicate the sample number, its objective (global analysis, organics, micro components, etc.), nature of the water point (spring, shaft well, well, etc.), point of the collection (coordinates, tie–in with landmarks, etc.) and the specifics to be taken into account in the analysis. Besides time and date of sampling, name and position of the responsible person are mandatorily indicated.

2.2.6 Sample Safekeeping and Delivery to the Laboratory

Most manuals of ground water testing restrict time of sample safekeeping. Unfortunately, due to long distances between points of the field works and laboratory these recommendations are not always doable. Samples collected in the field are often delivered over thousands of kilometers and safe–kept in collections for months. Change of environment cannot affect their composition. In transport in winter, water samples may be frozen and glass vessels may crack. Besides, chemical analyses of preliminarily frozen samples are dubious as their original composition may not completely recover at melting. In such cases for mineral components, it is more reliable to use, not glass, but plastic vessels and for organic components, vessels of Teflon or stainless steel.

Samples are safe-kept and transported under conditions preventing the disruption of their hermeticity and change of waters composition. In particular, for sample conservation cooling or freezing is commonly applied, which is most effective immediately after the sample collection. Cooling of samples by melting ice or in a refrigerator at 2–5°C and safekeeping them in a dark place in most cases is sufficient for conservation of samples for several days. Freezing to –20°C and lower allows increasing the sample safekeeping period to months and longer. But in such cases it is necessary to use vessels from plastic (for instance, PVC) and to observe the freezing and melting procedure in order to obtain samples without changes. Samples in glass vessels and for microbiologic analysis are not frozen.

Lengthy safekeeping may noticeably affect the content of volatile and organic compounds. Samples of natural gas and water with high content of volatile components should be held stopper down.

At last, it is important not to lose labels with information about the samples. Labels may unglue, rub-off, wash-off, etc. It is especially important to preserve the sample number and information about its purpose. For this reason these parameters are recommended to be duplicated directly on the vessels in a felt pen.

For the control of possible sample contamination in safekeeping and transport sometimes ***travel blank samples*** are collected. They are only to control samples for volatile and unstable organic components. Such samples are prepared from distilled water. They are plugged, labeled and recorded in laboratory, then transported into the field and back into the laboratory together with test samples. They are used together with field blanks in order to differentiate sources of samples contamination at the late stages of testing. At least one travel blank sample must be prepared and analyzed for each group of samples for volatile organic compounds.

Hydrogeochemical testing is ended in laboratory at the moment of recording them in the laboratory registration ledger with mandatory indication of their number and volume.

3

Processing of Testing Results

Testing results in the final analysis are disparate numerical measurement data of the ground water with different property and composition parameters. At that, at each point of testing there may be several measurements of the same parameter. Some of them are attributed to work (test) samples, some others, to control samples. All these unprocessed measurements are ***observed values***. Their processing has the objective to obtain the most reliable idea of ground water properties and composition at individual testing points. Any testing point may have only one most likely value of each parameter, and this value is called ***measurement result***. Observed values and measurement results in combination determine two levels of the databank, based on which is built a model idea of hydrochemical state of the studied geologic object (Figure 3.1). The entire processing of testing results may be subdivided into three basic stages:

1. Processing and systematizing observed values.
2. Determination of measurements results and their reliability.
3. Modeling and visualization of analytical hydrogeochemical state.

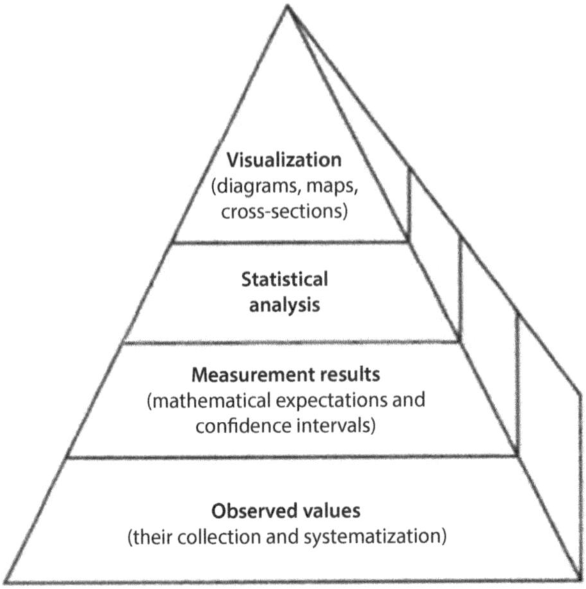

Figure 3.1 Testing results' processing procedure from collection of observed values to construction of maps and cross-sections.

3.1 Processing and Systematization of Observed Values

Observed values are all results of individual measurements of ground water sample property and composition parameters. These measurements are conducted either directly in the field or in water samples by different methods, different organizations, and often in different years. The large and most responsible part of the observed values are results of chemical analyses. These analyses describe the composition both of test samples and of control samples. These data require, first of all, careful checking and systematization.

3.1.1 Checking the Observed Values

Errors in analytical work and in copying the analyses results are unavoidable. Blunders may be the consequence of disorderly devices, poor quality chemicals or distilled water. Also possible are errors associated with calculations and typos by technicians. There are several ways to eliminate blunders in analysis results of macro components. Most of them relate to global complete analysis for macro-components. Following are the four most simple and accessible methods to discover errors.

1. From the balance of equivalent content. This technique is based on water neutrality in the electric charge. Due to this neutrality the sum of cation equivalents must be equal to the sum of anion equivalents. Neutral compounds are not included in the equivalent balance. Deviation from the balance may be expressed as absolute value, as the difference $\sum C_{eq,cat} - \sum C_{eq,an}$ ($\pm$meq·l^{-1}) or as a percent this difference makes of the entire sum:

$$\delta\% = 100\frac{\sum C_{eq,cat} - \sum C_{eq,an}}{\sum C_{eq,cat} + \sum C_{eq,an}}, \quad (1.71)$$

where δ% is charge balance error, $C_{э,cat}$ and $C_{э,an}$ are total contents of the cations and anions, respectively, in meq·l^{-1}.

In the neutral water H^+ and OH^- in equivalent balancee compensate one another. However, in very acid waters content of H^+ may exceed 1 meq·l^{-1} and affect the solution's value equivalent balance. For this reason in acid waters with the determined pH, the content of the hydrogen ion in meq·l^{-1} should be taken into account, which is equal:

$$C_{eq,H^+} = 1000 \cdot 10^{-pH}, \quad (1.72)$$

It should be included in the cation sum. This control is not applicable in the case of shortcut analyses. In good and complete global analysis this deviation must not exceed 1–2% at the salinity greater than 150 mg·l^{-1}. With the salinity decline the value of the acceptable errors increases to 3% at salinity around 100 mg·l^{-1} and to 5% for waters with salinity of up to 25 –30 mg·l^{-1}.

The chemical balance is a coarse estimate. It is unsuitable for the correctness estimate of the micro components content and content of compounds without electric charge. Values δ% usually have negative sign. Their great size indicates either a blunder in the determination or the presence of components unaccounted for by the analysis. Noticeable deviation may be caused by elevated water acidity, presence of polyvalent metal ions, and orthophosphates PO_4^{-3}, HPO_4^{-2}

and H_2PO^{-1}. But orthophosphate content (PO_4^{-3}, HPO_4^{-2} and H_2PO^{-1}) in water depends on pH value. The PO_4^{-3} ion may be encountered only at pH greater than 9.3 and practically does not affect equivalent balance of ground water. In the presence of phosphorus in water, relative HPO_4^{-2} and $H_2PO_4^{-1}$ concentrations may be calculated from pH valuee. Ions of polyvalent metals are present only in small amounts and rarely affect the ion balance noticeably.

2. From relationship between dry residue and salinity. Difference between the values of dry residue and total salinity usually does not exceed 3%. Doubts regarding the correctness and completeness of the analysis are acceptable if the difference exceeds this value. Reliability of this check depends mostly on the completeness of analysis, contents of organic compounds, bicarbonate and hydration of salts in the deposit. Presence of nonvolatile organic matter, incompleteness of the analysis and hydration of salts in the precipitate facilitate overstating the dry residue relative to the salinity. For this reason the calculated salinity usually is between the values of dry residue before and after its annealing. High bicarbonate content in fresh water, on the contrary, usually results in understatement of the dry residue.
3. By comparing specific electric conductivity and salinity. For most ground waters the specific electric conductivity in micro–siemens, multiplied by 0.65±0.1, is close to the value of the salinity in $mg \cdot l^{-1}$. This is not an exact correlation as the conductivity of a solution depends on its composition and salinity. The multiplier 0.65 is applicable only to relatively diluted solutions and usually increases when general salinity exceeds 2 – 3 $g \cdot l^{-1}$.
4. Comparing repeated analyses by typical ratios. Components' concentration ratios in the water are more conservative than the concentrations proper. For this reason changes in concentrations associated with errors are usually not accompanied by a change in the concentrations of other components. Drastic deviation of concentrations' ratio between two components may indicate the presence of substantial errors. Such may be the ratios of calcium and bicarbonate, sodium and chlorine and others, which are relatively stable in the studied hydrogeological object. When determined values substantially deflect from control ratios, an additional check

is called for. This technique of estimating analyses should be used reasonably as the stability of control ratios is relative.

The aforementioned control methods are suitable for the discovery of blunders only in the composition of mineral components. Nevertheless, when coordinating with analysts, they help either avoid analysis errors or discover the presence of components, which were not included in the analysis.

Analytical composition of natural gases is released by the laboratory in fractions (percent) of volume. The check is in the determination of the total content of all components and in the evaluation of possible sample contamination by air. In the global analysis the total content of components must be equal to 1 (100%). Noticeable deviation of the sum of all analytes from 1 indicates either incomplete analysis (some important component not analyzed) or a rough error (in calculations; or a typo). Sample contamination with air is possible to establish from the presence of O_2 and ratio of N_2 and Ar in its composition. The air could be captured in the field at sample collection, and in a device–analyzer when in the laboratory during the sample transfer. If the presence of air components in the tested water is impossible based on conditions of its occurrence or formation, the contamination is eliminated by way of recalculating analysis results (see above).

3.1.2 Systematizing the Observed Values

The observed values, even after their checking, describe not the composition and properties of the underground waters but the composition and properties of the water in a specific sample, and may have different dimensions. For mineral components, original concentrations are usually expressed in weight units relative to the solution, for gas components, in fractions relative to the natural gas volume, and for many organic compounds, in fractions of extractant's or concentrate's weight. For a complete description of underground water properties and composition at a specific water point these concentrations should be expressed in equal or comparable units, which allow the analysis of their interaction. In a case of gas components for this, it is necessary to have corresponding gas–water ratio values, and for organic compounds, concentrating or degree of extraction factors.

Five basic formats of expressing concentrations of components dissolved in ground waters may be identified: by the weight, by the volume, by the amount, by the energy and by their charge (Table 3.1). If

Table 3.1 Dimensions of concentration in hydrogeochemistry and their purpose.

Concept	Dimension	Units of dimension	Components
By the mass	Weight/volume	mg·l^{-1}, g·l^{-1}, g·kg^{-1}, w.%, ppm etc.	Any
By the volume	Volume/volume	ml·l^{-1}, m^{3}·m^{-3}, ml·kg^{-1}, v.% etc.	Gas and volatiles
By the electric charge	Number of charges/volume	meq·l^{-1}, geq·l^{-1},	Ions
	Fraction of charges	meq-%	
By the amount of components	Moles/volume (molarity)	mole·l^{-1}, mmole·l^{-1}, mole·kg^{-1}, mmole·kg^{-1}	Any
	Molar fractions	Fractions of units, percent	Any
	Moles/weight of solvent (molality)	mole·kg^{-1}, mmole·kg^{-1}	Any
By the energy	Partial pressure	Pa, H·m^{-2}, atm, at, bar, etc.	gas and volatile
	Relative thermodynamic concentrations (activity, fugitiveness)	Dimensionless, numerically associated with dimension of original concentrations	Any

the hydrochemical task is associated only with mineral components, the most convenient format for comparing their concentrations is equivalent (meq·l^{-1}, g-equiv·l^{-1}, epm, meq–%). It enables checking analyses by equivalent balance and solving many issues of electrolyte interaction. If hydrochemical issues are limited only to volatile components (gas and organic), the most convenient format of their comparison is partial pressure. But if for solution of the issue, it is necessary to deal with the interaction between mineral, gas and organic components, it is necessary to use their quantitative concentrations: molarity, molar fractions or molality. Thermodynamical and chemical calculations traditionally, as a rule, are performed in the values of molality as the most convenient and correct form of expressing concentrations.

Most observed values are expressed in numerical values, which must contain only accurate digits whose reliability is provided for by the limit of measurement, and only one doubtful digit. If the measurement limit is equal to 0.001, the last accurate digit will be the third decimal point. After it, only one digit is acceptable, the rest are discarded. Extra digits may cause confusion relative to the true accuracy. The last digit of the value is not absolutely accurate in all analyses but may be used with a good extent of confidence in most calculations. When discarding a digit equal or greater than 5, the preceding digit should be increased by 1.

Field measurement and analysis results of all samples from one testing point are reviewed as a complex, as the study results of one and the same underground water. This complex includes results of analyses of both test and control samples. For this reason, first of all, it is necessary to subdivide all observed values into test values, intended for the description of hydrochemical environment, and control values, intended for the description of the reliability of conducted measurements. Analyses of repeated control samples may be attributed to either test or control.

All observed values of test samples, recalculated per unit volume of water, are collected and systematized together in the spreadsheets. Currently numerous program packages exist, which may be used for the processing of hydrochemical data. Among them the most accessible are standard spreadsheets, for instance, *Excel, OpenOffice.org. Calc, Gnumeric, Access* and some others whose capabilities in special calculations and graphics are quite limited. It is possible to use specialized scientific programs like *MatLab*, *SigmaPlot*, *Origin*, etc., which are also not adjusted for specialized hydrochemical calculations. At last, there are special programs, which cover both the recording of hydrochemical data and their systematization and processing. Such databases are included practically with all programs associated with hydrogeochemical modeling.

Initially for one and the same testing point there may be several observed values of the same parameter. Such a population of measurements of individual water properties or composition parameters at one testing point is the ***statistical sample*** of observed values. Such a sample may involve measurement results of one parameter in test samples, in repeated control samples, and also in dividable samples sent to different laboratories. The representativity of such a statistical sample depends only on the number of these measurements.

The spreadsheets, beside the observed values, include a description of water points, their topographic and stratigraphic tie, depths of occurrence and some water property parameters. Analytical data are mandatorily accompanied by the name of the laboratory responsible for their quality or source (paper, book, report or sample number).

3.1.3 Control of Measurement Quality

Requirements for quality of the measurements directly depend on the liability. Three levels of control may be identified:

I. No control of measurement quality;
II. Quality control only of sample analyses;
III. Quality control of sample analyses and their collection.

I. First case: no requirements to the quality of measurement ground water property and composition values. Thus, the testing is not accompanied by either the collection of control samples or reliability control of obtained results.

II. Second case: associated with testing consumer qualities of ground water. All GOSTs devoted to water (P 51592–2000, P 51232–98, 27384–2002 and 24902–81) regulate measurement of only its consumer qualities, their requirements cover water property and composition parameters not in the reservoir but in laboratory conditions, i.e., actually in samples. For this reason these GOSTs, dealing with a measurement's quality control, disregard field sampling errors. Their requirements are limited by parameter measurement accuracy in samples. Such control is similar to within-the-laboratory control of an individual quantitative chemical analysis technique and in accordance with the requirements under MI 2335–95 "Internal quality control of quantitative chemical analysis results" is based on within-the-laboratory quality parameters measurements.

Therefore, this control of the observed values in this case is limited to operational control, under MI 2335–95 and GOST P 51232–98, which is conducted at confidence level 0.95. Control is performed relative to the error standards for three major parameters: repeatability (convergence) – d_i, reproducibility – D_i and accuracy – K_i. These standards are either applied to the utilized analysis technique or calculated from its error parameters.

Control of repeatability is conducted by comparing the results of N parallel measurements with the limit (standard) of repeatability (***repeatable error range***) d_i. If this standard is not available, it is determined as follows:

$$d_i = Q(n)\cdot\sigma_{i,repeat}, \tag{1.73}$$

where d_i is the limit (standard) of repeatability (repeatable error range), $\sigma_{i,repeat}$ is standard repeatability error of the analysis technique, and $Q(n)$ is critical range factor, which depends on confidence level and random values distribution law (Table 3.2). Repeatability of the repeated analyses results is considered satisfactory if the difference between end values (max – min) in one selection of repeated samples is less than the repeatable error range d_i, i.e.,

$$C_{i,\max} - C_{i,\min} \leq d_i. \quad (1.74)$$

Otherwise either analysis should be repeated or its technique changed.

Control of reproducibility is conducted by comparison of the difference between mean values of the parameter *i* with different techniques and the ***reproducible error range*** D_i. For operative internal control at the confidence level 0.95 the value of this range is equal to

$$D_i = Q(n) \cdot \sigma_{i,repr}, \quad (1.75)$$

where D_i is limit (standard) of reproducibility, $\sigma_{i,repr}$ is standard deviation of reproducibility of the analysis technique, $Q(n)$ is critical range factor, which depends on confidence level and random value distribution law (Table 3.2). Reproducibility of different testing techniques is considered satisfactory if the difference between end (max – min) mean parameter *i* values is less than D_i values, i.e.,

$$\overline{C}_{i,\max} - \overline{C}_{i,\min} \leq D_i. \quad (1.76)$$

In this case the results of different techniques may be combined and used together for calculating the measurement results. Otherwise the control should be repeated and the cause of significative difference between the results of different techniques should be sought for.

Table 3.2 Critical range factors **Q(*n*)** at confidence level 0.95.

n	Q(*n*)	*N*	Q(*n*)
2	2.77	12	4.6
3	3.31	13	4.7
4	3.63	14	4.7
5	3.86	15	4.8
6	4.0	16	4.8
7	4.2	17	4.9
8	4.3	18	4.9
9	4.4	19	5.0
10	4.5	20	5.0
11	4.6	21	5.0

Operative accuracy control is conducted using comparison samples or spiked-samples. For this the reference value of the comparison sample is subtracted from measurement results and the obtained difference θ_i is compared with the accuracy standard K_i at the assigned confidence level. At confidence level 0.95 standard K_i corresponds with the margin of error of the reference component concentration $\pm\Delta C_i$. At confidence level 0.90

$$K_i = \frac{1.64}{1.96}\Delta C_i = 0.84\Delta C_i, \tag{1.77}$$

where ΔC_i is global error in the determination of component i certified reference concentration. For operative analysis accuracy control of test samples a spiked–sample is also used . For this, first of all respective selected variances of the test samples proper and their spiked–samples are calculated, i.e., s_i^2 and $s_{i,sp}^2$. Then the accuracy standard at confidence level 0.90 is determined from equation:

$$K_i = 0.84\sqrt{s_i^2 + s_{i,sp}^2}\,. \tag{1.78}$$

Testing results accuracy is considered satisfactory if at the assigned confidence level the systematic error value is $\theta_i \leq K_i$. If the value exceeds the set accuracy standard K_i, testing is repeated or the cause of error is sought for.

III. Third case: it is associated with geochemical and ecological studies of underground water where it is treated as the object of geologic medium, and at the center of attention are its properties and composition under conditions of natural occurrence. Then the quality measurement parameters must include both analysis errors and the errors associated with underground water extraction and testing. Such parameters are calculated from results of control sample analyses. With a sufficient number of these samples it is possible to estimate also testing technique quality parameters as a whole, namely, its detection levels (DL) and limit of measurement (QL), repeatability, reproducibility, systematic error and accuracy.

3.1.3.1 Sensitivity of Testing Techniques

Testing technique sensitivity as a whole is determined from the results of blank field sample analysis. In these samples the analyte is absent from the beginning. For this reason, if in these samples is discovered its positive mean concentration, it determines the noise level, below which concentrations of the analyte cannot be detected, i.e., they are not significative. As blank samples by default have initial reference analyte concentration $C_i = 0$, the noise level is equal to $\overline{C}_{i,\text{blank}}$, and its random error is equal to standard deviation of this mean value, $s_{i,blank}$ (at sufficiently large number of testing $\sigma_{i,blank}$). It is customary to determine testing technique's detection level (DL) as a whole from the equation:

$$DL = \overline{C}_{i,\text{blank}} + 3\sigma_{i,\text{blank}}, \tag{1.79}$$

And limit of measurement from the equation:

$$QL = \overline{C}_{i,\text{blank}} + u\,\sigma_{i,\text{blank}}, \tag{1.80}$$

where u is the value assumed equal to 6 in Russia and 10 in the USA.

Beside the field blank samples, instrumental and transport samples may be collected. They enable the determination of DL and QL change at different testing stages. Laboratories regularly analyze their blank samples in order to control DL and QL for individual analysis techniques. Comparing DL and QL of the testing and of the analysis techniques allows estimating the relative role of methodological errors at different stages of studies. It is desirable for DL and QL of analysis techniques and of the testing as a whole to be comparable, i.e., one would not eclipse the other.

3.1.3.2 *Precision of Testing Results*

Precision as the description of the extent of measurement results closeness to one another has two end values, minimum under conditions of repeatability and maximum under conditions of reproducibility. The former evaluation is a description of the precision of only testing technique, the latter takes into account random errors caused by any deviations from this technique (change of instruments, chemicals, analytical laboratories, etc.).

3.1.3.2.1 Repeatability of Testing Results

Evaluation of the testing results repeatability as a whole is based on values of the selected variances of the repeated samples' analysis data in different testing objects. Let us assume that in each of $\overline{N}$ testing points there are N results of repeated testing with selected variances $s_{i,1}^2, s_{i,2}^2, \ldots, s_{i,n}^2$. First of all it is necessary to assure the uniformity of these selected variances. For this is used Q_t – criterion (Cochran test), which is equal to the fraction of the largest selected variance in the sum of all variances according to the equation:

$$Q_t = \frac{s_{i,\max}^2}{\sum_{n=1}^{\overline{N}} s_{i,n}^2} \tag{1.81}$$

where $s_{i,n}$ is selected variance at a specific testing point with sequential number n; if the total number of these points is N , $s_{i,max}$ is maximum standard deviation of repeated testing in one of these points.

The difference of maximum variance is recognized as significative if the calculated ratio Q_t is greater than its critical value in Table 3.3 at confidence level 0.95 and two degrees of freedom, $f_1 = N - 1$ and $f_2 = N(N-1)$. If $Q_t > Q_{t,table}$, the corresponding maximum variance is excluded from further calculations and the procedure is repeated for the next maximum – valuee variance, etc., until the calculated Q_t becomes less than or equal to $Q_{t,table}$. The remaining testing point variances are considered uniform, and from them the weighted average's standard deviation is estimated; it describes the repeatability of one testing technique at its numerous uses for the same

Table 3.3 Critical values of Cochran criterion $\mathbf{Q}_t$ at confidence probability 95%.

f_2	$f_1 = n - 1$				
	1	**2**	**3**	**4**	**5**
2	0.99	0.97	0.94	0.91	0.88
3	0.97	0.87	0.80	0.75	0.71
4	0.91	0.77	0.68	0.63	0.59
5	0.84	0.68	0.60	0.54	0.51
6	0.78	0.62	0.53	0.48	0.44
7	0.73	0.56	0.48	0.43	0.40
8	0.68	0.52	0.44	0.39	0.36
9	0.64	0.48	0.40	0.36	0.33
10	0.60	0.44	0.37	0.33	0.30
11	0.57	0.42	0.35	0.31	0.28
12	0.54	0.39	0.33	0.29	0.26
13	0.52	0.37	0.31	0.271	0.24
14	0.49	0.35	0.29	0.26	0.23
15	0.47	0.34	0.28	0.24	0.22
16	0.45	0.32	0.26	0.23	0.21
17	0.43	0.30	0.25	0.22	0.20
18	0.42	0.29	0.24	0.21	0.19
19	0.40	0.28	0.23	0.20	0.18
20	0.39	0.27	0.22	0.19	0.17
30	0.29	0.20	0.16	0.14	0.12

water. This standard deviation of testing technique's repeatability is estimated so:

$$S_{i,\text{repeat}} = \sqrt{\frac{\sum_{n=1}^{\overline{N}} s_{i,n}^2}{\overline{N}-1}}, \tag{1.82}$$

where $\overline{N}$ is the number of points with uniform variances. If $\overline{N}$ is great it is possible to assume that value $s_{i,repeat} \approx \sigma_{i,repeat.}$, i.e., it describes the repeatability of the entire testing technique as a whole including the errors of sample collection, preparation, transport and analysis. For this reason the testing repeatability value must be as close as possible to the valuee of within-laboratory repeatability but cannot be smaller than it.

3.1.3.2.2 Reproducibility of Testing Results

Reproducibility of testing results takes into account errors associated with any testing technique modifications (different conservation methods, different devices, laboratory, etc.). Usually hydrogeochemists are certain of the reliability of their testing technique and tend to doubt the job of analysts. For control of analysts' work they collect dividable samples and send them to different laboratories. For this reason reproducibility of testing results usually is associated with the change in analysis techniques but not testing techniques. For estimating the reproducibility in this case are used analysis results of dividable samples performed in different laboratories. First of all average analyte concentration $\overline{C}_{i,l}$ is calculated separately for each laboratory (or analysis technique). Then based on these averages the analyte's total average significance is determined:

$$\overline{C}_{i,L} = \frac{\sum_{l=1}^{L} \overline{C}_{i,l}}{L}, \tag{1.83}$$

where L is the number of laboratories (or techniques). Then the value of ***inter-laboratory variance*** (***between-laboratory variance***) is calculated as follows:

$$s_{i,L}^2 = \frac{\sum_{l=1}^{L}\left(\overline{C}_{i,l} - \sum_{l=1}^{L}\overline{C}_{i,L}\right)^2}{L-1}. \tag{1.84}$$

The reproducibility variance includes also inter-laboratory variance – $s^2_{i,L}$ and repetitiveness variance $s^2_{i,\text{repeat}}$. Thus, standard deviation of testing results' reproducibility is estimated from the following equation:

$$s_{i,repr} \approx \sqrt{s^2_{i,L} + s^2_{i,\text{repeat}}}. \tag{1.85}$$

Reliability of this standard deviation of reproducibility is higher if the number of dividable samples analyzed in comparable laboratories is higher. The value of the standard deviation's reproducibility must be as close as possible to the value of the repeatability standard deviation of testing results but cannot be smaller than it.

3.1.3.3 *Testing Correctness of the Results*

Such testing makes sense when there is a direct or indirect idea of the true value of measured parameter in reservoir conditions. Instead of the true parameter value is used some reference – $\hat{C}_i$. When it is impossible at the testing to preserve the component composition, the testing correctness is estimated based on elemental composition, namely, from its preservation in the process of extraction, collection, safekeeping and analysis of water samples. The measure of correctness is systematic error θ_i, i.e., mean value (mathematical expectation) of the difference between average measured value $\overline{C}_i$ and its reference analog $\hat{C}_i$:

$$\theta_i = \overline{C}_i - \hat{C}_i. \tag{1.86}$$

In hydrochemical testing the nature of reference values depends on the kind of control samples.

- In the case of blank control samples it is assumed that the analyte is absent and $\hat{C}_i = 0$;
- In the case of comparison samples (standard samples and certified mixes) the analyte concentration $\hat{C}_i$ is considered as the reference as long as it is certified by other more reliable measurement methods;
- In the case of spiked–samples reference value is determined by adding a specific amount of the analyte, as a result of which

$$C_i = C_{i,\text{spike}} \frac{V_{\text{spike}}}{V_{\text{s.}}}, \tag{1.87}$$

where $C_{i,\text{spike}}$ is the analyte concentration in added spiked–solution, V_{spike} is the volume of added spiked–solution and V_s is sample volume.

The measure of a testing technique's correctness is the systematic error.

3.1.3.4 *Systematic Error of the Testing Results*

The systematic error describes the value of mathematic expectation of observed values' deviation from the sought–for true value. Its measure is based on a comparison of parameters' measurements in samples with the reference value either in comparison samples or in spiked–samples.

Comparison sample (standard sample and certified mix) is used for the determination of a technique's systematic testing errors at the collection and preparation of water samples. For this purpose in the field are collected water samples with certified analyte concentration, i.e., they are prepared, safe–kept and transported to the laboratory together with work (test) samples. Then the reference value is the certified analyte concentration itself ($\hat{C}_i$), which is accompanied by the description of the total absolute error $\Delta\hat{C}_i$. Systematic error of the comparison sample is equal to:

$$\theta_i = \overline{C}_i - \hat{C}_i, \tag{1.88}$$

where $\overline{C}_i$ is mean analyte content *i* in the comparison sample from analysis results. To estimate the systematic error value – θ_i is used t_t **Student's test** whose value is:

$$t_t = \frac{|\theta_i|}{\sqrt{\dfrac{s_i^2}{N} + \dfrac{\Delta\hat{C}_i^2}{3}}}, \tag{1.89}$$

where s_i^2 is the variance relative to the $\overline{C}_i$ value, *N* is the number of collected comparison samples. If the calculated t_t is greater than in Table 3.4, the systematic deviation of the certified concentration is significative and it should be taken into account.

The spiked–sample or samples with additives also enable estimate of the systematic error at the sample collection and preparation. The analyte concentration in spiked–samples $\overline{C}_{i,\text{ст}}$.must be greater than in paired test samples $\overline{C}_i$ by the value of added analyte. Then the systematic error for spiked–samples is equal:

$$\theta_i = \left(\overline{C}_{i,\text{spike}} - \overline{C}_i\right) - \hat{C}_i = \left(\overline{C}_{i,\text{spike}} - \overline{C}_i\right) - \overline{C}_{i,\text{spike}}\frac{V_{\text{spike}}}{V_s}, \tag{1.90}$$

Table 3.4 Student's factor t_t at different confidence levels.

F		**1**	**2**	**3**	**4**	**5**	**6**	**7**	**8**	**9**	**10**	**15**	**20**	**30**
Confidence level	0.6	1.37	1.06	0.98	0.94	0.92	0.91	0.90	0.89	0.88	0.88	0.87	0.86	0.85
	0.7	1.96	1.39	1.25	1.19	1.16	1.13	1.12	1.11	1.10	1.09	1.07	1.06	1.05
	0.8	3.08	1.89	1.64	1.53	1.48	1.44	1.42	1.40	1.38	1.37	1.34	1.32	1.31
	0.9	6.31	2.92	2.35	2.13	2.02	1.94	1.90	1.86	1.83	1.81	1.75	1.72	1.70
	0.95	12.71	4.30	3.18	2.78	2.57	2.45	2.36	2.31	2.26	2.23	2.13	2.09	2.04

The value of these systematic errors is also estimated by Student's t–test. However, the t value in this case is calculated from the following equation:

$$t_t = \frac{|\theta_i|}{\sqrt{\frac{s_i^2}{N} + \frac{s_{i,\text{spike}}^2}{N} + \frac{\Delta\hat{C}_i^2}{3}}} \quad (1.91)$$

where s_i^2 and $s_{i,\text{spike}}^2$ are variances of mean values $\overline{C}_i$ and $\overline{C}_{i,\text{cп.}}$, respectively, and $\Delta\hat{C}_i$ is the global error of certified analyte addition's concentration value in the sample. If the calculated t_t is greater than in Table III–4, the systematic deviation is considered significative and it should be taken into account.

The significative systematic errors are corrected, i.e., added to the mathematical expectation of the parameter values with the sign always opposite to the error sign. If the systematic error is not significative, its value is disregarded.

3.1.3.5 *Testing Results' Accuracy*

Accuracy is the omnibus description of testing quality. It determines closeness extent of the analysis result to the sought–for true (or accepted reference) value. The accuracy measure of an individual measurement is ***global error***.

According to equation (I–67) global error is equal to the sum of systematic – θ_i and random – $\underline{\Delta}_i$ errors. Total absolute random error $\underline{\Delta}_i$ includes testing random errors, i.e., its reproducibility, values of θ_i, and depends on the confidence level.

For this reason global error of the testing technique is determined so:

$$\Delta C_i = \theta_i \pm \underline{\Delta}_i = \theta_i \pm u\sqrt{\underline{s}_i^2 + \underline{\underline{s}}_i^2} = \theta_i \pm u_{Si}, \quad (1.92)$$

where $\underline{s}_i^2$ and $\underline{\underline{s}}_i^2$ are variances of the systematic and random errors, respectively, u is quantile of normal distribution and s_i is the global error's standard deviation. At standard confidence level 0.95 we get:

$$\Delta C_i = \theta_i \pm 1.96 s_i. \quad (1.93)$$

If the value of systematic error is known and significative, it is subtracted from measurement results. The remaining error in this case is determined only by the value of global error's standard deviation:

$$\Delta C_i = 1.96 s_i. \quad (1.94)$$

The testing technique of global errors's standard deviation value s_i is the most important parameter of hydrogeochemical testing technique's quality. It allows evaluating the testing results' reliability as a whole. For this reason, the higher the liability for the reliability of a hydrogeochemical study, the higher are requirements to the substantiation of its value. With sufficient number of control samples the s_i value may be replaced by σ_i.

3.1.4 Measurements Results and Their Reliability

Observed values are only the source material used for the substantiation of ground water properties and composition parameter values. Properties and composition parameter values calculated based on the observed values are the ***measurement results***. If there is only one observed value, it is the measurement result. The greater the number of observed values used for the substantiation of measurement results, the higher its reliability. The determination of measurement results is regulated and under GOST **P** ISO 5725-1-2002 depends on the number (one or several) of observed values, their averaging method (arithmetic average value, median, standard deviation, etc.) and presentation format. Sometimes the introduction of standard corrections is required (for instance, bringing gas volume to normal temperature and pressure). Under GOST **P** 51592-2000 measurement results are described by two values: mathematical expectation and confidence interval, i.e., the format is $\overline{C_i} \pm \Delta_i$.

3.1.4.1 Mathematical Expectation

The mathematical expectation is most likely significance of sought-for true value. At a great enough number of observed values closest to mathematical expectation is ***mode*** of the selection. However, usually we deal with small selections of observed values, and the nature of their probability distribution is not known, which prevents sufficiently accurate estimate of mode. For this reason for estimation of the most likely concentrations is commonly used median, arithmetic average or geometric average. For this, any reliable data are used, which remain after the exclusion of observed values considered impossible or inconsistent with the other values of the selection, in other words, ***misses*** or ***outliers***.

Median value is a value positioned in the middle of an ascending-ordered results series. If the number of observed values N is odd, $\overline{C}_i$ is assumed equal to the middle term of the series and if it is even, to the arithmetic average of the two medium terms. The median is insensitive to end–point (drastically outstanding) measurement results and for this reason is sometimes convenient if the number of measurements is small ($N < 10$).

Arithmetic mean is used on the assumption of normal, symmetric probability distribution of the observed values. In this case

$$\overline{C_i} = \frac{1}{N}\sum_{n=1}^{N} C_{i,n}, \tag{1.95}$$

where $C_{i,n}$ is concentration of the component i in n^{th} sample, N is the number of observed values.

Geometric mean is used on the assumption of normal, symmetric distribution not of the observed values but their logarithms. Then

$$\overline{C_i} = \sqrt[N]{\prod_{n=1}^{N} C_{i,n}}. \tag{1.96}$$

The difference between the arithmetic and geometric mean usually depends on the value of random relative error. If this is small (less than 10%), this difference may be disregarded. However, with a large relative error geometric mean is more reliable. Most spreadsheets have special formulae for calculating these mean values.

3.1.4.2 Confidence Interval

Confidence interval describes the reliability of mathematic expectation, namely, range of the sought–for true value at the assigned confidence level. The positive and negative parts of this interval, i.e., lower and higher mathematic expectations, are called ***confidence limits.*** At symmetric error distribution relative mathematic expectation, when these limits are equal to one another, they are called ***confidence error***. Confidence interval value of mean values is usually expressed through the value of confidence error as $\pm\Delta_i$ and is found from the equation:

$$\pm\Delta_i = u\frac{\sigma_i}{\sqrt{N}}, \tag{1.97}$$

where u is quantile of normal distribution, N is number of observed values, and σ_i is standard deviation of global errors (at $N \geq 20$), i.e., the precision of the testing technique as a whole. The quantile value depends on the probability distribution and is used for the determination of the ***confidence level*** P. The value of this level describes the probability that the sought–for true value is within the limits of the given confidence interval. At $u=1$, $P=0.68$, at $u=1.96$, $P=0.95$, and at $u=3$, $P=0.997$.

Confidence error at confidence level $u=1$, i.e., at $P=0.68$ is equal:

$$\Delta_{st,i} = \frac{\pm\Delta_i}{u} = \frac{\sigma_i}{\sqrt{n}} \qquad (1.98)$$

and is called the ***standard error.***

In a case of lognormal distribution of observed values the standard deviation of global error describes, in essence, the mean of measured value relative deviation from average (by how many times, not by how many). For this reason confidence interval of geometric mean values describes the range of ***relative confidence error*** $\pm\delta_{i.}$ The absolute and relative confidence errors are associated through the following equation:

$$\pm\Delta_i = \pm\delta_i \cdot \overline{C}_i. \qquad (1.99)$$

Under GOSTs P ISO 5725-1-2002 and 27384-2002, confidence boundaries (lower and upper) at the confidence level 0.95 are considered to be the standard of measurement errors for water properties and composition parameters, i.e., the ***limiting errors***. In GOSTs these standards are usually expressed in percentages of the lower boundary of relative global error whose value increases with the concentration approach to the detection limit. For this reason standards of errors are sometimes given depending on the values of measured parameter (Table 3.5). Under these GOSTs measurement errors of natural, potable and waste water properties and composition must not exceed these standards.

At confidence level 0.95 confidence interval of ground water quality measurements depends only on the accuracy (global error) of the testing technique and the number of observed values:

$$\pm\Delta_i = 1.96\frac{\sigma_i}{\sqrt{N}} = \frac{\Delta C_i}{\sqrt{N}} = 1.96\underline{\Delta}_i. \qquad (1.100)$$

In MS Excel spreadsheets, for calculating the confidence interval at $N < 30$, it is possible to use the statistical function CONFIDENCE or procedure of descriptive statistics in Data analysis. In the former case it is necessary to introduce confidence level, i.e., signification (Alpha), selection's standard deviation and the number of observations (Size). In the descriptive statistics (Service, Data analysis) it is necessary to introduce the interval with observed values (Input interval) and to indicate the Confidence level (Level of reliability).

Table 3.5 Standards of relative measurement errors in water properties and composition under GOST 27384–2002.

Value of water quality parameter, mg/l	Standards of random relative error, ±δ%,	
	In natural and waste waters	In potable waters
< 0.000005		80
0.000005 – 0.00001	80	70
0.00001 – 0.0001	70	60
0.0001 – 0.001	60	50
0.001 – 0.01	50	40
0.01 – 0.1	40	30
0.1 – 1	35	25
1 – 10	30	20
10–100	25	15
100–500	20	
500–1000	15	

In the absence of measurement qualities' control and parameters, the confidence interval is determined regardless of the testing purpose from the value of the ***selection's standard deviation*** s_i, whose value is:

$$s_i = \sqrt{\frac{\sum\left(C_i - \bar{C}_i\right)^2}{N-1}}, \tag{1.101}$$

where N is the number of observed values. Obviously, for estimate of the confidence interval in this case, at least two testing events for each water point are necessary. As N increases, s_i value tends to σ_i.

For standard deviation of a small selection instead of quintal, u is used Student's factor t_t whose value depends on the degree of freedom equal N–1 and confidence level (see Table 3.4). For this reason in the absence of error parameters the confidence interval of measurement results is determined by equation:

$$\pm\Delta_i = t_t \frac{s_i}{\sqrt{N}}. \tag{1.102}$$

In the description of the confidence interval for water consumer quality parameters in equation 1.100 the values of within-the-laboratory global absolute error ΔC_i or standard deviation σ_i of the analysis technique are used. In the determination of the confidence interval for water quality parameters in the reservoir conditions the same parameters are used, but derived from control samples for the entire testing technique.

At the availability of only one measurement the confidence interval has the broadest range $\pm\Delta_i = 1.96\sigma_i$. The greater the number of repeated testing and observed values, the narrower the confidence interval. But its value may be substantially lowered only with 2, 3 or 4 observed values. Any further increase in the number of testing events is economically unreasonable.

Testing's global error allows estimating the optimum number of repeat testing events at one point for achievement of the required confidence interval. According to equation (I–100), at confidence level 0.95, minimum number of repeat testing events for achievement of the assigned confidence interval $\pm\Delta_i$ is:

$$N = \left|\frac{\Delta C_i}{\pm\Delta_i}\right|^2 \tag{1.103}$$

Measurement result values ($\overline{C}_i \pm \Delta_i$) are systematized in the form of relational **database**. Such a measurement result base is presented as a matrix, in which the lines correspond with testing objects and the columns, with parameters of these objects (their tie with landscape locations, environmental parameters, properties and composition). In spreadsheets of this database each testing object has only one row of averaged parameters. Such databases are available in practically all special hydrochemical programs (*AquaChem, The Geochemist's Workbench, GWB Essentials, HydroGto, SOXXXX, NETPATH, PHREEQC, PHREEQCI, PHRQPITZ, WATEQ4F, MINTEQA2*, etc.). The format, list of included components and parameter selection in such databases are usually quite individual. However, each testing object has coordinates, which enable construction of ***post maps*** (*source data maps*) and ***actual data maps***. Posted on these maps are all testing points with symbols indicating the type of water object (source, shaft well, well, etc.) and with the studied parameter values. The composite actual data map shows at each testing point the list of performed studies, i.e., the list of parameters, for which there are measurement results. These databases with the management system are the basis for a hydrogeochemical description of the tested hydrogeological object and modeling of hydrochemical processes.

3.2 Modeling of the Hydrogeochemical Condition

Measurement results are the basis for forming the comprehensive, most complete and reliable concept of ground water hydrochemical condition (state). Under ***hydrochemical condition*** in this case is understood the entire population of formalized correlations of ground water properties and composition parameters between themselves and with the time and space coordinates. This concept is based on the data of ground water analytical composition, and for this reason it is tentative. The comprehensive concept of ground water hydrochemical condition is possible only based on its genuine composition. Formation of this tentative concept is what determines the essence of ***modeling the hydrochemical condition***, which includes two stages:

- analysis and expressing formalized correlations between ground water property parameters and composition, i.e., empirical–statistical modeling and
- analysis and expressing distribution values of the same parameters in coordinates of studied time and space ranges, i.e., space–temporal modeling.

The modeling under consideration is based exceptionally on numerical data. Hydrogeochemists have to look for means of its visualization. This is manifested by the use of various schemes of expressing ground water composition and their comparison, of correlation diagrams, hydrochemical maps and cross–sections. However, the main task of this modeling is the search for unique determinative correlations, which would allow tying up the condition parameters between themselves and with the time and space coordinates. With complex geologic objects such functions are practically absent. All ties between ground water properties and composition parameters in time and space have properties of both deterministic and random correlations. For instance, water salinity is a function of a great number of factors. Some of them have random, non–unique influence (interaction with enclosing rocks, ion exchange, mixing with waters of a different composition, etc.), and some others have unique influence either only by increasing or only by decreasing the salinity (leaching of highly soluble rocks – limestone, gypsum, mixing with the sea water in the near–shore zone or with depths brines, desalinating by the atmospheric precipitation, etc.). As a result, the underground water salinity formation has both deterministic and random components.

The deterministic component is known as mean value (mathematical expectation), trend of the line or plain of equal values, etc. As a rule, it is the information about large–scale notable trends in parameter value changes and may be represented through specific deterministic functions. ***The random component*** is an unpredictable, complex and usually small-scale variability, about which too little is known in order to express it by a deterministic function but still enough to express it by a statistical function and probability parameters.

For this reason in the analysis of measurement results and construction of condition model the hydrogeochemist always deals with functions of two components: deterministic and random. The greater the role is of the deterministic component, the closer properties and composition parameters are associated between themselves and with space and time coordinates, the more specific and unique the model of state (condition) structure is, and the wider are opportunities for the interpolation and extrapolation. On the contrary, the greater the role is of the random component, the weaker are functional correlations, the less determined is the structure of hydrochemical state, and the fewer are opportunities for interpolation and extrapolation techniques.

3.2.1 Empirical-statistical Modeling

In a substance, the ***empirical-statistical simulation*** is the search for the ground water composition's equation of state. In this case such equation is a population of any empirical correlations between the values of ground water properties and composition parameters. These correlations enable the ability to estimate value of a parameter from value of a different parameter by formal statistical aggregates, and together they determine the content of an analytical hydrochemical state's ***empirical statistical model***. Description of such a model includes a characteristic of the statistical distribution of the measured parameter values and a description of their interrelation.

3.2.1.1 Anomalies and Background

As a rule, the description of a statistical distribution begins with the analysis of distribution of individual parameter values. For this purpose histograms or curves are constructed of differential and integral frequency distribution. If values' measurements results comply with the normal distribution, for their description there may be used traditional methods of statistical description of the data selection. However, high-resolution ground water studies indicate that the concentrations of chemical components in most cases are distributed independently and not normally. This creates certain difficulties in their statistical analysis. In such cases either some functions of measurement results (logarithmic, exponential, etc.) are used, which provide for the normal distribution, or methods of nonparametric statistics.

A general statistical description of a selection measurement result for one parameter must include the ***range***, i.e., difference between maximum and minimum values, ***mean value*** $-\overline{\overline{C}}_i$, ***standard deviation*** $-\overline{\sigma}_i$ or ***variation factor*** $\overline{\delta}_i$.

At normal distribution, average parameter values are calculated using the following equation:

$$\overline{\overline{C}}_i = \frac{1}{\overline{N}} \sum_{n-1}^{\overline{N}} \overline{C}_{i,n} \tag{1.104}$$

at lognormal distribution, from equation:

$$\overline{\overline{C}}_i = \sqrt[\overline{N}]{\prod_{n-1}^{\overline{N}} \overline{C}_{i,n}} \tag{1.105}$$

where $\overline{C}_{i,n}$ is concentration of a component i at the water-point j, $\overline{N}$ is number of testing points.

Standard deviation of a selection is determined from the normal distribution equations:

$$\overline{s}_i = \sqrt{\frac{\sum_{n-1}^{\overline{N}} \left(\overline{C}_i - \overline{\overline{C}}_i\right)^2}{\overline{N} - 1}} \tag{1.106}$$

and for the lognormal:

$$\overline{\overline{s}}_i = \sqrt[\overline{N}]{\frac{\sum_{n-1}^{\overline{N}} \left(\lg \overline{C}_i - \lg \overline{\overline{C}}_i\right)^2}{\overline{N} - 1}} \tag{1.107}$$

Variation factor is determined as the ratio of standard deviation to mean value.

Statistical description of ground water properties and composition measurement distribution results allows the subdivision of their values into background and anomalous. The ***background*** measurement results are those, typical of normal conditions and especially common within the extents of the given geologic object or part thereof (water horizons, complexes, blocks, etc.). It is believed that local anomalous factors (field, contamination focus, fault, etc.) do not take part in the formation of their value. In their totality the background properties and composition parameters are a description of the hydrogeochemical background, i.e., typical ground water properties and composition of a given study object. For a reliable description of background parameters sometimes water–points are specialy tested under normal hydrogeological conditions at sufficient distance from fields, contamination areals, faults, etc.

On the opposite end against the background are ***anomalous values*** of water properties and composition parameters, which have limited, local development and form under influence of relatively rarely occurring factors. For separating background and anomalous values a notion of the ***background range*** is used, which describes the range of background values. This range is determined from the following equation:

$$\pm\Delta\overline{\overline{C}}_i = \overline{\overline{C}}_i \pm u\bar{s}_i, \tag{1.108}$$

where u is quantile whose value depends on the confidence level. B.A. Kolotov et al. (1983) determine the background for hydrochemical search of ore fields at $u = 3$, i.e., on condition that the probability of encountering anomalous value is only 0.3%.

3.2.1.2 *Water Distinction in Quality Parameters*

Ground water properties and composition change constantly and gradually. For this reason ground water does not have specific boundaries, which would separate them by the quality. Usually the closer the testing points to one another, the smaller the difference in their water quality. At very close location of the points the difference may become less than the confidence interval. In such a case waters of the tested objects are actually indistinguishable by quality. The issue is: when do waters of different objects really differ from one another in quality, and where should the boundaries between waters of different quality be placed?

Statistical analysis provides objective criteria for differentiating ground waters by their quality parameters. Let us assume that at numerous ground water testing events in two different points were obtained two different concentrations of one and the same analyte *i*, $\overline{C}_{i,1}$ и $\overline{C}_{i,2}$. The difference between these values is considered really existing and is called ***significant difference*** if its value ($\overline{C}_{i,1} - \overline{C}_{i,2}$) exceeds its error at the assigned confidence level.

Standard deviation of this difference σ_Δ, if using the testing technique with equal standard deviation value of the global errors σ_i, is equal to:

$$\sigma_\Delta = \sqrt{\frac{\sigma_i^2}{N_1} + \frac{\sigma_i^2}{N_2}} = \sigma_i\sqrt{\frac{1}{N_1} + \frac{1}{N_2}} \tag{1.109}$$

With normal distribution of the $\left|\overline{C}_{i,1}-\overline{C}_{i,2}\right|$ difference values and with its standard deviation σ_Δ, the quantile value is determined as follows:

$$u=\frac{\left|\overline{C}_{i,1}-\overline{C}_{i,2}\right|}{\sigma_\Delta}=\frac{\left|\overline{C}_{i,1}-\overline{C}_{i,2}\right|}{\sigma_i}\sqrt{\frac{1}{N_1}+\frac{1}{N_2}}. \tag{1.110}$$

At the confidence level 0.95 the quantile should not exceed 1.96. For this reason, if the u value is greater than the Table values at the assigned confidence level, for instance, if

$$\overline{C}_{i,1}-\overline{C}_{i,2}>1.96\sigma_i\sqrt{\frac{1}{N_1}+\frac{1}{N_2}}=\Delta C_i\sqrt{\frac{1}{N_1}+\frac{1}{N_2}}, \tag{1.111}$$

the distinction of the measured results is significative.

This equation indicates that the greater the number of repeated testing events under the same technique, the more detailed the study results are. At the confidence level 0.95 and $N_1=N_2=N$ the analyte concentrations at different testing points are distinguished significatively if:

$$\text{at } N=1,\ \overline{C}_{i,1}-\overline{C}_{i,2}>2.77\sigma_i=1.41\Delta C_i;$$
$$\text{at } N=2,\ \overline{C}_{i,1}-\overline{C}_{i,2}>1.96\sigma_i=\Delta C_i;$$
$$\text{at } N=3,\ \overline{C}_{i,1}-\overline{C}_{i,2}>1.60\sigma_i=0.82\Delta C_i$$

In the absence of the data about the global error of the testing technique as a whole (or of its standard deviation) the significativeness of the differences between the ground water properties and composition measurement results is determined using the Student's criterion .

If two results of analyte concentration measurement $\overline{C}_{i,1}$ and $\overline{C}_{i,2}$ have the selection variance $s_{i,1}^2$ and $s_{i,2}^2$, then the standard deviation of their difference will be determined from the following equation:

$$s_\Delta=\sqrt{\frac{s_{i,1}^2}{N_1}+\frac{s_{i,2}^2}{N_2}}. \tag{1.112}$$

At small selection volumes the difference of measurement results $\left|\overline{C}_{i,1}-\overline{C}_{i,2}\right|$ complies with Student's t_t-distribution. For this reason:

$$t_t=\frac{\left|\overline{C}_{i,1}-\overline{C}_{i,2}\right|}{s_\Delta}=\frac{\left|\overline{C}_{i,1}-\overline{C}_{i,2}\right|}{\sqrt{\frac{s_{i,1}^2}{N_1}+\frac{s_{i,2}^2}{N_2}}}. \tag{1.113}$$

The difference between the values of compared measurement results is significative if the calculated value t is greater than in Table 3.4 at the assigned confidence level and degree of freedom:

$$f=\left[\frac{s_{i,1}^{2}}{N_{1}}+\frac{s_{i,1}^{2}}{N_{2}}\right]^{2}\cdot\left[\frac{\left(\frac{s_{i,1}^{2}}{N_{1}}\right)^{2}}{N_{1}-1}+\frac{\left(\frac{s_{i,1}^{2}}{N_{2}}\right)^{2}}{N_{2}-1}\right]^{-2}, \quad (1.114)$$

where f is the degree of freedom rounded to a whole integer.

If the dispersion ratio,

$$F_{t}=\frac{s_{i,1}^{2}}{s_{i,2}^{2}}, \quad (1.115)$$

i.e., Fisher's test/criterion, is less than in Table 3.6 at the degrees of freedom N_1-1 and N_2-1 and confidence level 0.95, then it is possible to assume that $s_{i,1}^{2}=s_{i,2}^{2}=s_{i,}^{2}$. Then the degree of freedom $f = N_1+N_2-2$, and

$$t=\frac{\left|\bar{C}_{i,1}-\bar{C}_{i,2}\right|}{s_{i}}\sqrt{\frac{N_{1}N_{2}}{N_{1}+N_{2}}}. \quad (1.116)$$

At last, if the rule $N_1= N_2= N$ is observed, the degree of freedom $f=2(N - 1)$, and

$$t=\frac{\left|\bar{C}_{i,1}-\bar{C}_{i,2}\right|}{s_{i}}\sqrt{\frac{N}{2}}. \quad (1.117)$$

If ground water testing was conducted by one technique with close values of variances s_i^2 only 2 times at each point ($N_1=N_2=2$), mean parameters are distinguishable on condition

$$\overline{C_{1i}}-\overline{C_{2i}}>4.3s_{i};$$

if 3 times ($N_1=N_2=3$), then

$$\overline{C_{1i}}-\overline{C_{2i}}>s_{i}\frac{2.78}{\sqrt{1.5}}=2.27s_{i};$$

Table 3.6 Values of Fisher's criterion F_t for confidence level 0.95 (f_1 is the number of degrees of freedom of a larger variance, f_2 is the number of degrees of freedom of a smaller variance).

	f_1									
f_2	**1**	**2**	**3**	**4**	**5**	**6**	**7**	**8**	**9**	**10**
1	161.45	199.50	215.71	224.58	230.16	233.99	236.77	238.88	240.54	241.88
2	18.51	19.00	19.16	19.25	19.30	19.33	19.35	19.37	19.38	19.40
3	10.13	9.55	9.28	9.12	9.01	8.94	8.89	8.85	8.81	8.79
4	7.71	6.94	6.59	6.39	6.26	6.16	6.09	6.04	6.00	5.96
5	6.61	5.79	5.41	5.19	5.05	4.95	4.88	4.82	4.77	4.74
6	5.99	5.14	4.76	4.53	4.39	4.28	4.21	4.15	4.10	4.06
7	5.59	4.74	4.35	4.12	3.97	3.87	3.79	3.73	3.68	3.64
8	5.32	4.46	4.07	3.84	3.69	3.58	3.50	3.44	3.39	3.35
9	5.12	4.26	3.86	3.63	3.48	3.37	3.29	3.23	3.18	3.14
10	4.96	4.10	3.71	3.48	3.33	3.22	3.14	3.07	3.02	2.98

If 4 times, then

$$\overline{C_{1i}} - \overline{C_{2i}} > s_i \frac{2.45}{\sqrt{2.0}} = 1.73 s_i .$$

The result is that it is impossible to compare ground water quality only by mean values of their measurement results. It is necessary to account for their measurement error, i.e., values of standard deviations or confidence interval. Waters with all indistinguishable mean quality parameters are impossible to consider different. Distinction by the quality is acceptable only for those waters which have significative differences in the properties or composition.

3.2.1.3 Search for the Factors

The model of hydrochemical state/condition includes also correlations of water properties and composition parameters between themselves (internal factors) and with the geologic environment – external factors, which are possible to observe and measure. The analysis of these correlations begins with cluster analysis, ends with factor analysis and includes dispersion, correlation and regressive analyses. A detailed description of these statistical techniques is beyond the scope of this textbook. We'll restrict ourselves to the description of their purpose and role in the generation of hydrochemical state/condition model.

Cluster analysis (*data clustering*). This is the umbrella name for a large selection of algorithms used for the generation of classifications. Sometimes it is considered as a technique for the classification and partition. In this case its role is partitioning the entire testing point selection into groups – clusters, in which the waters are similar between themselves based on measurements results. The technique begins with the generation of elementary clusters or classes, which then are grouped by the attributes of parameter similarity. As a result, clusters may be sequentially gathered into ever expanding groups, which form a hierarchical tree (arboreal/tree-like clustering). The attributes may be quantitative (depths of occurrence, time, coordinate points, concentration, salinity, pH, etc.) or qualitative (water point type, water horizon, reservoir lithology or stratigraphy, etc.). The first, non–statistical step in the cluster analysis may be perhaps considered the existing ground water hydrochemical classifications, which will be reviewed below. Cluster analyses do not have means of statistical verification but they enable optimization of database management and its preparation to subsequent statistical analyses.

Dispersion (variance) analysis, anova is analysis of the effect of qualitative attributes (climate, lithology, etc.) or their combinations on water properties and composition values in order to identify factors of ground water formation. It is assumed that the factors are known and it is just a matter of estimating how substantial (significative) their influence is. Depending on the number of these affecting factors, single–factor, dual–factor, etc., variance analyses are distinguished. This analysis is based on the study of significative distinctions between the means caused by various factors. Actually the variance of these mean values is compared with variance of random unaccounted–for factors and their difference is estimated with ***Fisher's criterion***. If the influence is significative, the factor works and has to be taken into account in the model generation. The use of such analysis is possible if the compared selection values are independent from one another and have normal distribution. Program MS Excel accepts the use of procedure "Anova" in Data analysis.

Correlation analysis determines the probability of a deterministic linear correlation between values of any two parameters. The objective of the correlation analysis is to determine if there is a correlation and how strong it is. The mathematical description of this strength is the ***correlation coefficient*** (*correlation factor*) $\overline{R}$. Correlation factors between all selected research properties and composition parameter pairs form the ***correlation matrix***. Values of the correlation factors are non-dimensional and range between −1 (strict inverse linear correlation) and +1 (strict direct linear correlation). They are a measure of relative role played by the deterministic and random components. The closer the correlation factor value to −1 or to +1, the higher the role of functional (deterministic) correlation. Used for numerical variables is Pirson's linear correlation factor. Its significativeness depends on the selection volume and is determined from the ***Student's criterion*** at a given confidence level. Usually it is believed that when the value of its modulus is less than 0.4 the deterministic linear correlation is not established, when the values are 0.6–0.8 a functional bond is present and with the value greater than 0.95 it is dominant.

Regression analysis determines the nature of functional correlation (linear, polynomial, exponential, logarithmic, etc.) between parameter values. This analysis is usually applied when the correlation is significative. It is usually determined by selecting a mathematical model – the type of a regression equation, at which the correlation has maximum significance of the ***determination factor*** ($\overline{R}$*-squared, coefficient of determination*) $\overline{R}^2$. The determination factor is the main attribute of a regression model and determines with what accuracy this model describes (approximates) the initial data. The maximum approximation is reached by selecting such trend function (regression equation) at which the determination factor $\overline{R}^2$ is closer to 1. Search of factors (constants) unknown ahead of time in the regression equation is performed with the ***least squares method***. Significativeness of the distinction of these factor values from 0 is checked with ***Student's criterion***. And significativeness of the model, i.e., the reliability of $\overline{R}^2$ values, is checked with ***Fisher criterion*** F_t.

The regression analysis of a correlation between two parameters is easily conducted in *Excel*. For this purpose, based on results of measuring parameters (two columns) is constructed a scatter chart. Program MS Excel accepts the use of procedure "Regression" in Data analysis.

In a case when functional correlation is suggested between parameters of water properties and composition, values of the ratios of several parameters (Na/Cl, Cl/Br, He/Ar, etc.) are directly statistically analyzed with the determination of mathematical expectation and variance values. The smaller the value of the ratio variance, the higher the probability of the direct correlation between the compared components.

Factor analysis is a technique of the search for optimal system of substantial or operative *factors* affecting the entire population of studied attributes in order to construct a ***factor model*** (*factor pattern*). In particular, this technique deals with causes affecting values of underground water quality parameters. Factor analysis is based on studies of the correlation matrix and variance values of each parameter. It is directed toward shrinking of the *number* of variables (data reduction), i.e., merging them in factors and determining the structure of interconnection between them (classification of variables). At that, variables strongly correlated between themselves are merged and those non–associated between themselves are treated as different factors. The identification of the real natural causes causing the merging of parameters into some factor, naturally, is up to the researcher. The structure of interrelation determines mutual subordination and nature of the bonds between geologic environment parameters and underground water chemical composition. Eventually the simplest and most usable model of factor analysis is formatted as a system of linear equations:

$$Y_1=k_{1,1}F_1 + k_{1,2}F_2 + k_{1,3}F_3 + \ldots + k_{1,p}F_p + U_1$$

$$Y_2=k_{2,1}F_1 + k_{2,2}F_2 + k_{2,3}F_3 + \ldots + k_{2,p}F_p + U_2$$

$$\ldots$$

$$Y_n=k_{n,1}F_1 + k_{n,2}F_2 + k_{n,3}F_3 + \ldots + k_{n,p}F_p + k_nU_n,$$

where Y_1, Y_2 … Y_n are variable parameters of the water's chemical composition, F_1, F_2 … F_p are factors affecting these parameters, $k_{i,j}$ are ***loading factors***, i.e., significance measures of the effect of a given factor on the variability of a given parameter and $U_1, U_2 \ldots U_n$ are ***unique factors*** determining the residual variance (scatter values) of a given parameter, which cannot be determined even by the entire multitude of the identified factors. The factor analysis method is directed to the estimation of loading factors and specific factors.

The correlations of the empirical–statistical mathematical model describe the internal structure of the hydrochemical state/condition of a geologic object, which allows determining unmeasured ground water properties and composition parameters based on the measured ones. Quite a number of special statistical program packages (*STADIA, SAS, STATISTICA, SPSS for Windows, Eviews,* etc.) exist for analysis of these correlations. The best and most used among them is the integrated system of data analysis and management *STADIA STATISTICA*, which includes actually all necessary statistical analytical methods.

3.2.2 Space–time Modeling

Distribution scheme of ground water quality parameter values in space–time coordinates is a different model. This model, as a rule, cannot be described by a single equation or a system of equations, which would uniquely determine values of these parameters as functions of time and space coordinates. For this reason such a model is formally constructed only based on measurement results and coordinates of their tie–in and is called ***spatial–temporal schematic model***. The degree of detail and reliability of such a model are both directly dependent on the tested geologic object density and uniformity but its pitfalls are associated mostly with the lack of data for unsampled points. For minimizing these pitfalls interpolation and extrapolation methods are used, which, in their turn, require the knowledge/concept of functional correlation between the quality parameter values and space or time coordinates. These concepts are called ***metamodels*** or ***response surface***. Metamodels in this case are the instrument of approximation of a formally constructed schematic model to real ground water properties and composition distribution within the extent of the sampled geologic object.

The simplest linear expression of such metamodel determines the concentration $C_{i,0}$ at the point x_0 from two known concentrations $C_{i,1}$ and $C_{i,2}$ located at the points x_1 and x_2 (or at the moments of time t_1 and t_2). In this case it is possible to express the metamodel by a linear interpolation equation:

$$C_{i,0} = \bar{C}_{i,1} + \frac{\bar{C}_{i,2} - \bar{C}_{i,1}}{x_2 - x_1}\left(x_0 - x_1\right). \quad (1.118)$$

However, the C_i vs. x (or t) correlation is not always linear. In actuality this function is not known and is not permanent. For this reason more correct is the equation:

$$C_{i,0} = f(x) \pm \overline{\sigma}_i(x) \tag{1.119}$$

where $f(x)$ is an unknown correlation function of concentration vs. distance (or time), which is not necessarily linear, and $\overline{\sigma}_i(x)$ is the value of the error associated with uncertainty of this function and errors of the used measurement results. This equation is the basis of almost all correlation metamodels of water properties and composition parameters vs. space and time coordinates. It includes both the deterministic component $f(x)$ and random component $\overline{\sigma}_i(x)$. The first addend uniquely determines the concentration $C_{i,0}$ at the point x_0 as a function of any measurement result $C_{i,n}$ and distance $f(x)$. The second addend describes the statistical component, namely, a random error associated with uncertainty of function $f(x)$ nature, with testing errors, etc. Variable x in equation 1.119 is a representation of the coordinates of the unsampled point (x, y), which is often called in Russian publications the ***regionalized variable.***

If $\overline{N}$ testing points are located around the point x_0 and the function $f(x)$ is known, it is possible to forecast $\overline{N}$ values of $C_{i,0}$. However, these values will differ from one another by the amounts, which determine the nature of $\overline{\sigma}_i(x)$. Underground water, being a solution, is a system with distributed parameters. This means that its quality parameters change in time and space coordinates continuously and smoothly, with some gradient. For this reason there are no abrupt insurmountable boundaries in ground waters, like between minerals in rocks. With the approach of the surrounding testing points to the point x_0 differences in properties and composition of their waters decrease. When these testing points come together at the point x_0, the result is numerously repeated testing of a single object. But even in this case, testing results may differ within the value limits of testing's global error, which at confidence level 0.95 is equal to $\Delta C_i = 1.96\sigma_i$. Thus, with the decrease of distance between x_0 and any sampled point x_n in space (or in time), the value of the function $f(x)$ tends to $C_{i,n}$ and the value $\overline{\sigma}_i(x)$ decreases to the value of the testing global error ΔC_i. If the number of sampled points is $\overline{N}$, with their approach to the point x_0, the $C_{i,0}$ value in equation 1.119 tends to the mathematical expectation:

$$\underset{x \to 0}{C_{i,0}} \to \frac{1}{\overline{N}} \sum_{n=1}^{\overline{N}} C_{i,n} \pm \overline{s}_i(x),$$

where $C_{i,n}$ are the measurement results of parameter i at $\overline{N}$ testing points with sequential numbers n, and $\overline{s}_i(x)$ is standard deviation of the selection of these measurement results as a function of distance.

On the contrary, with distancing of testing points from the point x_0 the uncertainty value of functions $f(x)$ increases, the fraction of the deterministic component in equation 1.118 decreases and the fraction of the random component $\overline{s_i}(x)$ grows. This is expressed in that with an increase in distance between testing points the bond between the values of the same parameter weakens, and the value of their correlation factor decreases. The nature of the correlation between the deterministic and random components in equationя1.119 vs. the distance between testing points determines the content of metamodels and is estimated by geostatistical methods.

Geostatstics is a discipline in statistics, which studies and analyzes statistical distribution and interconnection between variables in space and time coordinates. Geostatistical methods enable estimating distances, at which values of a parameter are interrelated and their interpolation is acceptable.

At the basis of geostatistics is analysis of interrelations between the value pairs of the same parameter at two different points of space or time. Under the monitoring regime the position of these testing points is expressed through time t; under regional studies, through x and y coordinates; and at studies in 3D space, through x, y and z coordinates. The simplest geostatistical metamodel is the linear parameter self-correlation called autocorrelation. More complex metamodels describe parameter self-correlation in 2D or 3D space, and are expressed with the help of autocorrelograms, semivariograms or variograms.

3.2.2.1 Autocorrelation Metamodels

Autocorrelation is correlation between two values of the same parameter at the points separated by time interval Δt or distance Δx, which is called ***lag***.

We will familiarize with autocorrelation on the example of analysis of temporal series changes in concentrations under the monitoring regime. Let us assume that concentration component i was measured over the interval of time Δt. For the analysis of autocorrelation functions by means of Excel we'll input in 2 columns of the spreadsheet results of measuring time t and the corresponding concentration component i. These columns form ***temporal series*** (***time series***). They enable plotting of the correlation graphs concentrations vs. time. Similarly, we can plot graphs of concentration vs. distance. Usually values of the observed parameters are entered on the abscissa axis (x) and time (distance), on the ordinates axis (y) as shown in the left columne in Figure 3.2. Statistical description of such a series includes its mean value (mathematical expectation), for instance arithmetic mean value:

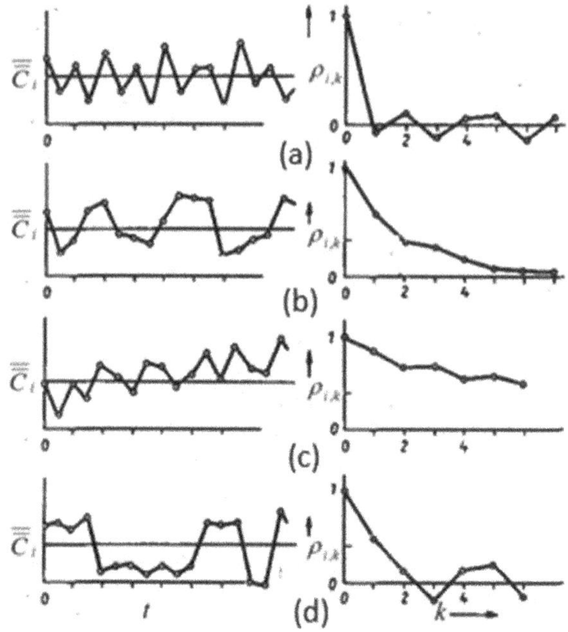

Figure 3.2 Temporal series and their corresponding autocorrelograms; a are independent measurements; *б* are dependent measurements; *c* are changes with drift; *d* are noisy periodicities.

$$\overline{\overline{C}}_i = \frac{1}{\overline{N}}\sum_{n=1}^{\overline{N}}\overline{C}_{i,n}, \qquad (1.120)$$

and the corresponding standard deviation:

$$\overline{s}_i^2 = \frac{1}{\overline{N}}\sum_{n=1}^{\overline{N}}\left(\overline{C}_{i,n} - \overline{\overline{C}}_i\right)^2, \qquad (1.121)$$

where $\overline{C}_{i,t}$ is concentration measurement results at the points separated by time (or distance) intervals with sequential numbers n, and $\overline{N}$ is the number of sampling points.

Autocorrelation of this temporal series describes the correlation between the series of values $\overline{C}_{i,n}$ measured at times $t_1, t_2, t_3, \ldots, t_n$, and the series of values measured at times $t_1+\Delta t$, $t_2+\Delta t$, $t_3+\Delta t$, …, $t_n+\Delta t$, where Δt is a constant value. Similarly, it is possible to compare the concentrations measured at the points $x_1, x_2, x_3, \ldots, x_n$, and at the points $x_1+\Delta x$, $x_2+\Delta x$, $x_3+\Delta x$, …, $x_n+\Delta x$, where Δx is a constant value. For the introduction of the second series we will copy the first column of concentrations and enter it next, to the right of it. The result is two equal data columns with equal values $\overline{\overline{C}}_i$, $\overline{\sigma}_i^2$ and number of pairs $\overline{N}$. The lag value Δt (or Δx) between the values in these columns is initially equal to 0, and the ***self-correlation coefficient/factor*** R_0 between them is equal to 1. Now we'll move the right column one cell down. Then next to one another will be the concentrations measured over time interval Δt (or space interval Δx), and the number of these pairs will shrink to $\overline{N}-1$. Now the values of compared pairs are not equal to one another and the correlation factor $\overline{R}_1$ between them will be less than 1. If the column moves one more cell down, the lag value will increase to $2\Delta t$ (or $2\Delta x$), and the number of pairs will decrease to $\overline{N}-2$, then the correlation factor, most likely, will become even smaller. Let us assign the sequential number k to each move of the right column down. Then the lag between the values of the compared pairs will be equal to $k{\cdot}\Delta t$ (or $k{\cdot}\Delta x$) and the number of the compared pairs will be $\overline{N}-k$. If each measurement result has the number n, it will be juxtaposed by a measurement result of the same series with the number $n+k$ and the ***autocorrelation factor*** between them will be determined from the following equation:

$$\overline{R}_{i,k}=\frac{1}{\overline{N}-k}\sum_{n=1+k}^{\overline{N}-k}\frac{\left(\overline{C}_{i,n}-\overline{\overline{C}}_i\right)\left(\overline{C}_{n+k}-\overline{\overline{C}}_i\right)}{\overline{s}_i^2}. \qquad (1.122)$$

However, the basis of the reviewed metamodel is not autocorrelation per se but its correlation with the lag values. The graph of this correlation $\overline{R}_{i,k}$ vs. lag values $k{\cdot}\Delta t$, plotted from the measurement results, is called *sample correlogram* or ***autocorrelogram***. The examples are shown in the right column of Figure 3.2.

Selected autocorrelograms give the idea of correlation function of the autocorrelation factor between the measurement results of one parameter vs. lag values, i.e., vs. time or distance interval. Such function is called ***autocorrelation function (ACF)***. Its dimensionless value at $k=0$ is always equal to either +1 or −1. For this reason this function should be dealt with only at $k>1$. Its real nature is unknown and is substituted by such metamodel function, which best corresponds with the selected autocorrelogram. As metamodels are the most commonly used functions of purely random process, first and second order autoregression, sliding average, etc.

A special place among them belongs to the metamodel of ***purely random process***, which is often called ***white noise***. This metamodel is viewed as a sequence of absolutely unconnected but equally normally distributed measurement result values. According to this metamodel, autocorrelation factor values have significance 0 *npu* $k>1$ and variance $1/n$, where n is the term number in the series. It is describing that very noise, which is unavoidably present in any hydrogeochemical testing and is describing the $\sigma_i(x)$ value at $f(x)=0$ in equation I–119. At quite significant number of measurement results and confidence level 0.95 the values of autocorrelation factor for this metamodel are within confidence interval $\pm 2/\sqrt{N}$. For this reason the metamodel of purely random process (white noise) should be used only for parameters, for which autocorrelation factor values are within this range. It means that no notable correlation of measured parameter vs. time or distance is discovered. On the contrary, autocorrelation factors are considered significative and suitable for interpolation and extrapolation only when they are outside the bounds of this interval. Thereby they determine predictability of the reviewed value series as a whole.

When the determinative correlation of measured parameter vs. time or distance is significative, it is desirable to review its format. As a rule, with the increase of lag value $k \cdot \Delta t$, the value of autocorrelation factor declines and approaches 0 (Figure 3.2, right–hand diagrams). This means that with the increase of time duration or distance the correlation between measurement results of the same parameter decreases. The time interval or distance, within which autocorrelation factors are significative and the interpolation is acceptable, is called the ***characteristic time*** or ***range effect***. Autocorrelation factors outside the bounds of this range do not differ significatively from 0, therefore, they do not include the deterministic component. Thus, it follows that the autocorrelation function and its metamodel make sense only within the bounds of range effect. For this reason, for substantiated interpolation and extrapolation the sampling should also be conducted with the time or distance interval smaller than the value of this range.

Obviously, all factors defining the type of a correlogram are possible to subdivide into two groups: random and correlatable. The random factors are $\bar{\sigma}_i(x)$, i.e., white noise of uncontrolled subordinate processes. Correlatable factors define the determination function $f(x)$ in equation I–119. Factors of these groups compete between themselves. At the low effect of correlatable factors, the random dominate, and all autocorrelation factors at $k > 1$ are not distinguishable from 0 (Figure 3.2a). The greater the influence of correlatable factors, the higher the role of autocorrelation functions, the greater the distance of its effect (Figure 3.2b).

Functional (deterministic) correlation of an autocorrelation function in a general case may be considered as a sum of two components: the function of a stable global growth or decline (Figure 3.2c) called the ***trend***, and a periodic function. Periodicity indicates the presence of unstable factors whose effect noticeably changes in time (seasonal changes, periodic discharge, etc.), and is manifested by regular change of the measurement results and autocorrelation factors sign (+ or –) (Figure 3.2d). It usually occurs only in the temporal series where seasonal or multiannual climatic fluctuations play a large role. In space it occurs extremely rarely, and the deviations from the trend usually are considered to be noise. The trend of water quality parameters changes with time, is manifested in the increase of the range effect, and is formed under influence of unidirectional and stable factors (climate warming, pollutant migration from its source, etc.). For instance, Figure 3.2c presents a correlation, for which the autocorrelation factor does not approach 0 in the entire range of lags.

Special significance in the analysis of temporal and spatial series has the identification of a trend as functional correlation of parameter values vs. coordinates of time and space. In geology this operation is often called ***trend analysis***. It makes sense if the deterministic component is significative. Trend analysis often leverages smoothing with sliding averages and regression analysis. Seasonal fluctuations are analyzed by special techniques of analytical seasonal smoothing and seasonal autoregression. When measurement errors are very large, a technique of negative exponentially weighted smoothing is used. Such smoothing is necessary for filtering out the noise and converting the correlation into a relatively smooth curve. Sometimes a functional correlation derived in the regression analysis by the least square technique is used as the smoothing line. In this case it is possible to derive trend water composition parameters correlation vs. two, three and even all four space–temporal variables. However, the trend correlations vs. three and four variables currently appear extremely rarely.

Metamodels of the autocorrelation functions enable linking together variance and measurement results, autocorrelation factors, and considering this link as a function not only of the time Δt but also of the distance Δx between sampling points. Thereby they enable differentiating and analyzing factors affecting the ground water quality and forecasting changes of these qualities in coordinates of time and space. But these autocorrelation metamodels may be used only in studies of linear changes in water properties and composition of the landscape. For instance, they may be applied for analysis of water quality changes with depth in a separate well or lengthwise a single line of testing along the section, filtration, etc.

Methods of computing autocorrelation functions, constructing correlograms and their analysis may be found in programs intended for analysis of temporal series, in particular, in the *MS Excel* suspension "Autocorrelation function" or *XLSTAT*, *STATISTICA*, '*Caterpillar*'-SSA, *EViews*, etc. The *MS Excel* suspension does not have the installer. It is simply copied into its suspension directory. The correlograms are recommended to plot when at least 50 measurement results are available. Unreliable or doubtful data cannot be used. If the data are insufficient, autocorrelation factors will be overvalued.

3.2.2.2 *Semivariance Metamodels*

However, under a regional testing the nature of the autocorrelation functions depends not only on the distance between sampling points but also on their position on the landscape. This correlation becomes ever more complex with the increase in anisotropism of the sampled medium. That is why at the self-correlation analysis of the parameter value on the landscape it is necessary to take into account not only the distancee between the sampling points but also its direction.

On a horizontal plain the distance between two testing points with coordinates x_1, y_1 and x_2, y_2 is a vector with the length:

$$\Delta x = \sqrt{(x_2 - x_1)^2 + (y_2 - y_1)^2}, \quad (1.123)$$

and with the deflection angle from x coordinate of:

$$\Theta = a\ \tan\left[\frac{y_2 - y_1}{x_2 - x_1}\right]. \quad (1.124)$$

For the analysis of self-correlation between the values of one parameter at arbitrarily placed testing points in a nonlinear space is used not autocorrelation but semivariance.

Semivariance in this case is understood as half of the variance between any pairs of values in the selection, i.e., half of the squared difference between two measurements of one parameter.

For the sake of simplicity we will first review measurement results for a series of testing points positioned along one line, assuming that their x coordinate is always possible to replace by Δx, i.e., the population of coordinates x, y, z. Let us assume that the content of analyte i is defined at n testing points with coordinates x_1, x_2, …, x_n. If the concentration of this analyte at the point x_1 is equal to $C_{i,1}$ and at a distance Δx, $C_{i,2}$, then the semivariance values of the parameter i at the point x_2 relative to point x_1 may be determined as:

$$\gamma_i = \frac{1}{2}\left(C_{i,x} - C_{i,x+\Delta x}\right)^2. \quad (1.125)$$

The semivariance value depends on Δx values. We may guess that with distancing of the selection points from the point x_1, the semivariance value will be increasing. In order to estimate this correlation between γ_i and Δx, the testing points are joined in pairs whose general number is equal to $0.5\overline{N}\left(\overline{N}-1\right)$. The testing points in these pairs are positioned at arbitrary distances from one another. For this reason they are grouped by the distance with step $n \cdot \Delta x$, where n has values 1, 2, 3,..., $\overline{N}$. The result is $\overline{N}$ selections of the measurement result pairs positioned at distances Δx, $2\Delta x$, ..., $n\Delta x$ from the point x_0 with mean value $\overline{C}_{i,1}$, $\overline{C}_{i,2}$, $\overline{C}_{i,3}$, ..., $\overline{C}_{i,n}$. We will input these mean values as a column in the Excel spreadsheet and copy it into the column to the right. We will get two columns with the offset equal to 0 ($k = 0$). At this offset the value γ_i for mean values will be equal to 0. If we offset the right column by k cells relative to the left one, the difference between mean values of the adjacent cells and valuc γ_i will increase. In this case the mean value of the parameter i's semivariance will depend on the number of offsets, i.e., k value:

$$\gamma_{i,k} = \frac{1}{2\left(\overline{N}-k\right)} \sum_{n=1+k}^{\overline{N}-k} \left(\overline{C}_{i,n} - \overline{C}_{i,n+k}\right)^2, \tag{1.126}$$

where $\overline{C}_{i,n}$ and $\overline{C}_{i,n+k}$ are parameter i mean values at the testing points, separated by the distance $k \cdot \Delta x$, $\overline{N}$ is total number of distance intervals, k is the ratio of mean distance between the compared testing points to Δx. It follows from this equation that:

$$2\gamma_{i,k} = \frac{1}{2\left(\overline{N}-k\right)} \sum_{n=1+k}^{\overline{N}-k} \left(\overline{C}_{i,n} - \overline{C}_{i,n+k}\right)^2 = \frac{1}{\left(\overline{N}-k\right)} \sum_{n=1+k}^{\overline{N}-k} \overline{C}_{i,n}^2 + \frac{1}{\left(\overline{N}-k\right)} \sum_{n=1+k}^{\overline{N}-k} \overline{C}_{i,n+k}^2 - \frac{2}{\left(\overline{N}-k\right)} \sum_{n=1+k}^{\overline{N}-k} \left(\overline{C}_{i,n} - \overline{C}_{i,n+k}\right) \tag{1.127}$$

The averaged value $2\gamma_{i,k}$ is called ***variance function***, $\gamma_{i,k}$ – ***semivariance function*** or simply semivariance. If we plot γ_{ik} along the Y axis and mean values of their corresponding distances along the X axis, we will get the diagram ***variogram***. If we substitute $\gamma_{i,k}$ values for 2 $\gamma_{i,k}$ in this diagram, we will obtain the ***semivariogram***. The nature of the curves will not conceptually change because of this substitution. For this reason the terms semivariogram and variogram are sometimes treated as synonyms.

As the compared variation series of one parameter are interval ones and have equal intervals, their variance may be calculated from the mathematic

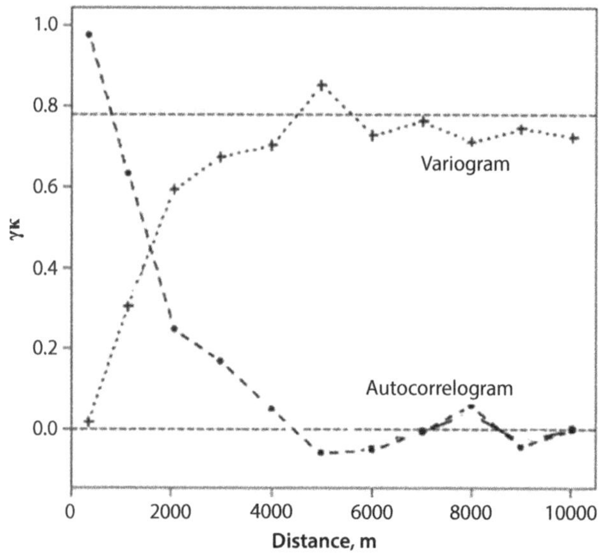

Figure 3.3 Comparing of the variogram and autocorrelogram for one selection.

expectation equal to 0. At that, we may assume that the value of these variances within the boundaries of the studied object does not change. In which case equation 1.127 may be formatted with values of variances and autocorrelation factors of the individual selections:

$$2\gamma_{i,k} = \overline{\sigma}^2_{i,k} + \overline{\sigma}^2_{i,k} - 2\overline{\sigma}^2_{i,k}\overline{R}_{i,k} = 2\overline{\sigma}^2_{i,k}\left(1 - \overline{R}_{i,k}\right), \qquad (1.128)$$

where $\overline{\sigma}^2_{i,k}$ is the variance value of parameter i and $\overline{R}_{i,k}$ is autocorrelation factor. For this reason, when the autocorrelation factor $\overline{R}_{i,k}$ approaches 1, the semivariance value $\gamma_{i,k}$ tends to 0 (Figure 3.3). And vice versa, when the autocorrelation factor is equal to 0, the semivariance is equal to the selection's variance.

Variograms are generated in order to determine the distance of reach of the deterministic correlation $f(x)$ of ground water properties and composition parameters depending on the distance and the effect of random factors $\sigma_i(x)$ in equation 1.119. The obtained experimental selection variogram gives an idea of some function, which is a "mirror image" of the autocorrelogram. However, it is impossible to use until the equation of this functional correlation of $\overline{R}_{i,k}$ or $\overline{\sigma}^2_{i,k}$ values vs. the lag size $k \cdot \Delta x$, which matches the experimental data best, is determined. This equation is a metamodel, which would best determine correlation of the semivariance factor vs. distance in landscape coordinates. Search for this equation boils down to the selection of such a function (Table 3.7), which best corresponds with the point distribution in the experimental variogram and allows calculating the $\gamma_{i,k}$ value. These metamodels are subdivided into stationary, nonstationary and combination.

Semivariance metamodels are used to describe statistical structure of individual parameters' value distribution in the landscape coordinates. Theoretically, at $k = 0$, i.e., at repeat sampling in one point, the semivariance factor $\gamma_{i,k}$ of any model must be equal to 0 and autocorrelation factor, to 1. Quite often repeat testing at the same point gives noticeably different measurement results of the same parameter, i.e., at $k = 0$ $\gamma_{i,k}$ is noticeably greater than 0. The value of such deviation is called the ***nugget effect*** (Figure 3.4). The cause of this may be measurement errors at testing or incorrectly selected distance between sampling points (spatial variations occur at closer distance than the selected distance). The name of this parameter is associated with appraisal of gold fields where "top–cut grades" of the metal content in samples were often unpredictable.

Stationary metamodels display the growth limit for the values of semivariance $\gamma_{i,k}$. This growth slows down with increasing lag and stops upon having reached some limit called ***variogram sill*** (Figure 3.4). The upper value of this sill usually corresponds with the flattening level of the variogram curve with the distance from the coordinate origin. In connection with this, stationary metamodels have the ***range effect***, i.e., the distancee, at which the $\gamma_{i,k}$ changes its values. At a distance smaller than the range effect, the forecast, interpolation and extrapolation are possible with certain probability. At distances greater than the range effect, such forecast is unacceptable. The values of the range effect and variogram sill determine the boundary separating zones of prevailing deterministic function $f(x)$ on the left and random unpredictable function with variance $\sigma_i(x)$ on the right. Together, the variogram sill and nugget effect determine the total variogram interval, which is equal to the selection's variance. Stationary models include metamodel of the nugget effect, Gauss, exponential and spherical. Most commonly used in hydrogeology is the exponential metamodel .

Nonstationary metamodels describe the variograms not having the growth limit to semivariance $\gamma_{i,k}$ values. They are used when the variogram sill is not established. Examples of these metamodels may be exponential and linear metamodels. Combination metamodels assume the use of two different metamodels together, for instance, linear or exponential ones, with the nugget effect, etc.

If the medium is isotropic and the distribution of parameter i value does not depend on the direction of the distance vector, the nature of the variogram also does not depend on the orientation of this vector. A variogram,

Table 3.7 Most commonly used variogram models.

Model	Parameters	Conditions	Functions
Stationary metamodels of variograms			
Nugget effect	$\bar{\sigma}^2_{i,\text{nug.}}$ – nugget variance, $\delta(k)$ – Kronecker delta equal to 1 if $k = 0$ and 0 if $k > 0$.	$\delta(k)>0$	$\bar{R}_{i,k} = \bar{\sigma}^2_{i,\text{nug.}}\delta(k) = 0$ $\gamma_{i,k} = \bar{\sigma}^2_{i,\text{nug.}}\left[1-\delta(k)\right] = \bar{\sigma}^2_{i,\text{nug.}}$
		$\delta(k)=0$	$\bar{R}_{i,k} = \bar{\sigma}^2_{i,\text{nug.}}\delta(k) = \bar{\sigma}^2_{\text{nug.}}$ $\gamma_{i,k} = \bar{\sigma}^2_{i,\text{nug.}}\left[1-\delta(\text{k})\right] = 0$
Limited linear (Bounded linear)	a is the value inverse to range effect; $\bar{\sigma}^2_{i,sill.}$ is sill variance.	$At\ k\Delta x \leq 1/\text{a}$	$\bar{R}_{i,k} = 1 - ak\Delta x$ $\gamma_{i,k} = \bar{\sigma}^2_{i,sill} ak\Delta x$
		$At\ k\Delta x > 1/\text{a}$	$\gamma_{i,k} = \bar{\sigma}^2_{i,sill}$
Exponential		$3/a$	$\bar{R}_{i,k} = \exp(-ak\Delta x)$ $\gamma_{i,k} = \bar{\sigma}^2_{i,sill}\left[1-\exp(-ak\Delta x)\right]$
Gaussian		1.73/a	$\bar{R}_{i,k} = \exp\left[-(ak\Delta x)^2\right]$ $\gamma_{i,k} = \bar{\sigma}^2_{i,sill}\left\{1-\exp\left[-(ak\Delta x)^2\right]\right\}$

Spherical		0≤kΔx≤1/a	$\overline{R}_{i,k} = 1 - 1.5ak\Delta x - 0.5(ak\Delta x)^3$ $\gamma_{i,k} = \overline{\sigma}^2_{i,sill}\left(1.5ak\Delta x - 0.5(ak\Delta x)^3\right)$
		kΔx>1/a	$\overline{R}_{i,k} = 0$ $\gamma_{i,k} = \overline{\sigma}^2_{i,sill}$
Nonstatic variogram metamodels			
Exponential	α and β –proportionality factors.	0<β<2	$\gamma_{i,k} = \alpha(k\Delta x)^\beta$

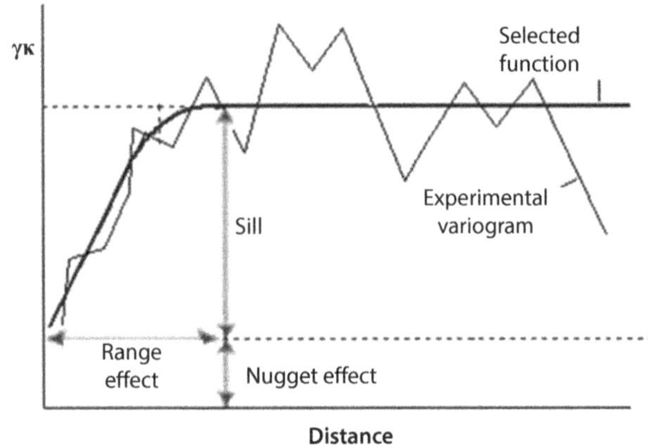

Figure 3.4 Ideal variogram and its parameters.

which depends only on the distance between sampling points, is called ***isotropic variogram***. Such variogram, plotted for a single point in all directions (in γ_k, x and y coordinates), forms a true round funnel. A variogram, which does depend on the direction of the distance between testing points, is called ***anisotropic variogram***. In a horizontally anisotropic medium the variogram funnel loses its round form and gets extended in the direction of greater orderliness. At that, the range effect often changes with equal value of the variogram sill.

Variogram metamodel is at the basis of any geostatistical study. It allows for the determination of spatial correlation between the data at arbitrarily placed testing points, for more precise planning the distribution of these testing points and for the evaluation of error estimate parameters at unsampled points. It describes statistical structure of water quality parameter values distribution in coordinates of the studied geologic objects, which is very important for interpolation while mapping ground water quality.

Currently practically every program package for processing the hard rock geology information includes means for computing 3D variograms for the assigned directions. They all have similar algorithms and require similar sets of source data. Among domestic petroleum programs, which perform geostatistical analysis, should be mentioned PO "BASPRO Optima" by the company group BASPRO in Tyumen. But most universal and popular in hydrogeochemistry is a geoinformation system *Golden Software Surfer* by Golden Software, which is intended for analysis of nonuniformly distributed regional information and regional mapping. Its instruments enable the selection of an optimal variogram model accounting for its isotropism.

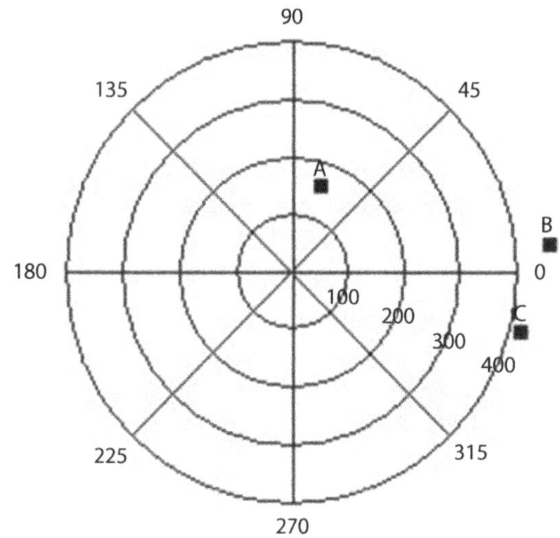

Figure 3.5 Variogram grid and its tolerance regions.

For saving the machine time and the amount of memory the variogram construction is performed not by a simple search of all comparison pairs but by their classification by lag value and direction. The ***angle tolerance***, i.e., acceptable deviation from the mean angle, and ***lag tolerance***, i.e., acceptable deviation from the mean lag, determine the ***tolerance regions***. These tolerance regions may be visualized as the special polar ***variogram grid*** shown in Figure 3.5. The program is moving the center of the variogram grid from one point to the next, sorting the pairs by their belonging to tolerance regions and summing up their squared difference $\left(C_{i,x} - C_{i,x+\Delta x}\right)^2$. The quotient from dividing the obtained total by the doubled number of pairs gives mean value of the semivariogram in each tolerance region. By tying these values with mean angle and lag values, the program generates variograms of different regional direction and analyzes them.

3.3 Classification and Visualization of Hydrogeochemical Parameters

Visually, ground waters are practically indistinguishable. The descriptions of their properties and composition are based only on their numerical values. From the very first steps in ground water study attempts were made to shrink the number of these numerical values and make them visual. For this, first of all, attempts were made to classify natural waters by properties and composition and also to express their variety graphically in order to compare these waters between themselves and to demonstrate changes in their properties and composition in time and space.

3.3.1 Chemical Classification of Ground Waters

Primeval ground water classifications by properties did not take into account the composition and were based mostly on taste. The waters were subdivided into fresh, brackish, salt, sour, bitter and sweet. A genuine ground water diversity was discovered in studies of their composition by methods of analytical chemistry. It turned out that waters equal in their taste qualities and physical properties may be substantially different in their chemical properties. Moreover, ground water composition became the main criterion of their qualities.

A first classification by the composition was based on the relationships of macro components. The water was differentiated by dominating cations and anions. However, it was found that the ground water composition may be dominated by strong acids and alkali but their properties to a much greater extent depended on the content of weak alkali and acids. For this reason in the names of water types were introduced second and even third in concentration cations and anions. Such formal ground water differentiation by the equivalent content of individual ions resulted in the emergence of a large number of excessively detailed, bulky and user–unfriendly chemical classifications. To this group belonged the classifications by S.A. Shchukarev (49 types), N.I. Tolstikhin (100 types), O.S. Dzhikiya (625 types), etc. In this connection were even undertaken attempts to number ground waters by the composition. Eventually these classifications gave way to simple and convenient formula expressing the water composition, which in 1928 proposed a physician and balneologist Mikhail Georgiyevich Kurlov (1859–1932). ***Kurlov's formula*** was a fraction whose numerator included anions and their content in descending order, and the denominator, cations in the same order. Concentrations were given in meq-% with values of at least 5%. In front of the fraction after the letter M followed the salinity, and behind the fraction, additional physical or chemical water properties (temperature, pH, Eh etc.). Sometimes in front of the fraction was indicated the content of additional components (mineral or gas).

$$CO_2\,0.012, M0.20\frac{HCO_3 33SO_4 15}{Ca27Mg12Na11}pH6.8; t = 10°C. \quad (1.129)$$

This formula was a compact expression of the general water composition in the most visual format.

It was found in brine studies that their properties substantially depended on the content of weak acids and alkali. So, in the process of studying brines in underground waters and lakes there emerged another, salt direction in ground water classification. This direction was based not on ion content but on the content of salts, which the waters could form at evaporation. First, it was established that the composition of these salts depended not on the water salinity and not on the concentrations of its individual ions but on the interactions of these ions between themselves, which was expressed in the salt composition. Second, it was noted that the leading role was played not by strong acids and alkali, which were often present in high concentrations, but by weak acids and alkali whose concentrations might be negligible because they first precipitated into deposit. The first on this way was Ch. Palmer, after them followed N.S. Kurakov, M.G. Valyashko, V.A. Sulin and O.A. Alekin.

Palmer's classification was proposed in 1911. It was based on studies of oil field brines and currently has only historical value. Ch. Palmer subdivided all macro-components dissolved in water by their acid-alkali properties into 5 groups. He subdivided the cations into 3 groups: strong alkali – Na^+ and K^+; weak bases (alkali-earth) – Ca^{2+} and Mg^{2+} and very weak bases (heavy metals) – Fe^{2+}, Cu^{2+}, and Al^{3+}. He subdivided the anions into 2 groups: strong acids – Cl^- and SO_4^{2-} and weak acids –HCO_3^-, CO_3^{2-}, etc. He came to a conclusion that ground water chemical properties were determined not by the contents of individual ions but by the composition of the salts, which they were capable of forming. Based on this he identified six groups of salts:

- Three groups by salinity, formed by strong acids: 1. first salinity (with strong alkali), 2. second salinity (with weak bases) and 3. third salinity (with very weak bases).
- Three groups of different alkalinity, by weak acids: 1. first alkalinity (with strong alkali), 2. second alkalinity (with weak bases) and 3. third alkalinity (with very weak bases).

Depending on the ratio of strong acid equivalent concentrations, on the one hand, and strong and weak bases, on the other, Ch. Palmer identified 5 ground water classes. He attributed to the 1st class the waters, in which there were less strong acids than strong alkali ($Cl^- + SO_4^{2-} < Na^+ + K^+$). The natural feature of these waters was the presence of first alkalinity salts, in particular sodium hydrocarbonate ($NaHCO_3$), and the absence of the second and third salinity salts, i.e., of the irremovable hardness. These were soft alkali waters associated with crystalline rocks. He attributed to the 2nd class the waters, in which the content of strong acids and strong alkali was equal ($Cl^- + SO^{2-}_4 = Na^+ + K^+$). And in the 3rd class he included the waters, in which strong acids were more than strong alkali but less than the sum of strong and weak bases ($Na^+ + K^+ + Ca^{2+} + Mg^{2+} > Cl^- + SO^{2-}_4 > Na^+ + K^+$). A typical feature of these waters was the presence of the second salinity salts and the absence of the first alkalinity salts. These hard waters are mostly associated with sedimentary rocks. In the 4th class were included the waters, in which the content of strong acids was equal to total content of strong and weak bases ($Na^+ + K^+ + Ca^{2+} + Mg^{2+} = Cl^- + SO^{2-}_4$). This class included the waters similar in their composition to the sea water. At last, Ch. Palmer attributed to the 5th class the waters in which there were more strong acids than strong and weak bases together ($Na^+ + K^+ + Ca^{2+} + Mg^{2+} < Cl^- + SO^{2-}_4$). Such waters with elevated contents of heavy metals were encountered relatively rarely in ore fields. Natural for them was the presence of the third salinity salts.

A major pitfall in Palmer's classification is merging into one group components different in their properties (chlorides with sulphates or calcium with magnesium) and the identification of types based on the equality of concentrations (the second and fourth types). Due to this the identified ground water classes are not equivalent in the distribution, some are practically absent, and the others are too massive. Besides, Ch. Palmer's classification is inconvenient in usage due to complexity of the water class determination from analysis data. In order to simplify this operation, J. Rogers in 1917 proposed a graphic technique, which was called ***the Palmer–Rogers diagram***. This technique is a diagram of a relative anion and cation content in the form of two parallel columns. At the base of the columns are shown the contents of strongest acids and bases, higher are shown weak and very weak ones (or the other way around). It makes it easy to determine the relative content of salts with different salinity and alkalinity, therefore, water class (see Figure 3.6).

In Russia a hydrochemical classification of ground waters based on their salt composition was being developed since 1917. N.S. Kurakov and M.G. Valyashko proposed it for salt lakes, V.A. Sulin, for deep underground brines, and O.A. Alekin, for onland surface waters. Most popular among them became the classification by V.A. Sulin.

Sulin's classification was proposed in 1946 mostly for oil field ground waters. At the basis of Sulin's classification is the same principle as Ch.

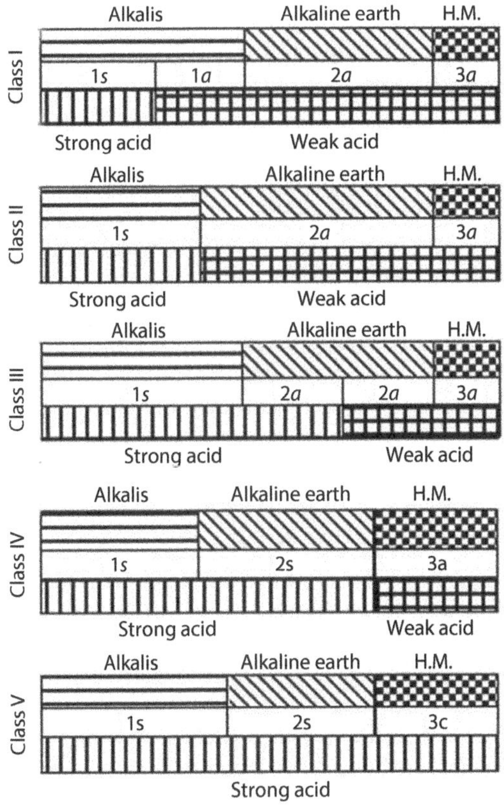

Figure 3.6 Palmer–Rogers diagrams reflecting all five water classes after Palmer. Here: Alkali – Na^{+}+K^{+}; Alkaline earth – Ca^{2+}+Mg^{2+}; H.M. – Fe^{2+}, Cu^{2+}, Al^{3+} and other heavy metals; Strong acid– Cl^{-}+SO_4^{2-}; Weak acid –HCO_3– + CO_3^{2-}; *s* – salinity, *a*– alkalinity.

Palmer's. But he expanded the circle of included macro-components to six: Cl^{-}, $SO^{2-}_{4}$$^{-}$, HCO^{-}_{3}, Na^{+}, Mg^{2+} and Ca^{2+}, and introduced very convenient ratios for comparing their concentrations:

$$\frac{\left[Na^{+}-Cl^{-}\right]}{\left[SO_4^{2-}\right]} \text{and} \frac{\left[Cl^{-}-Na^{+}\right]}{\left[Mg^{2+}\right]}.$$

V.A. Sulin identified four ground water types, which he associated with their formation conditions.

He attributed to the first type the waters, in which Na^+ and K^+ content (strong alkali after Ch. Palmer) was greater than total Cl^- and SO_4^{2-} contents (strong acids after Ch. Palmer), i.e.,

$$\frac{\left[Na^+ - Cl^-\right]}{\left[SO_4^{2-}\right]} > 1.$$

Actually this water type after V.A. Sulin corresponds with the 1st class of water after Ch. Palmer. He called these waters ***hydrocarbonate-sodium*** as only they are capable of forming from concentrating a salt, $NaHCO_3$.

He included the second type of waters, in which the Na^+ and K^+ content (strong alkali after Ch. Palmer) is higher than Cl^- but lower than total Cl^- and SO_4^{2-} content (strong acids after Ch. Palmer), i.e.,

$$\frac{\left[Na^+ - Cl^-\right]}{\left[SO_4^{2-}\right]} < 1.$$

V.A. Sulin called this type ***sulfate-sodium*** as these waters were capable of forming Na_2SO_4 salt.

The third type after V.A. Sulin includes waters, in which Cl^- content is higher than Na^+, K^+ and Mg^{2+}, i.e.,

$$\frac{\left[Cl^- - Na^+\right]}{\left[Mg^{2+}\right]} > 1.$$

This type, after the natural salt, which only these waters are capable of forming, he called ***chloride-calcium.***

At last, the last, fourth type after V.A. Sulin combines waters, in which the content of Cl^- is greater than Na^+ and K^+ but smaller than total content of Na^+, K^+ and Mg^{2+}, i.e.,

$$\frac{\left[Cl^- - Na^+\right]}{\left[Mg^{2+}\right]} > 1.$$

He called this type after the composition of a natural salt, which only these waters can form, ***chloride - magnesium.***

Thus, V.A. Sulin noticeably simplified the water type determination technique by proposing relatively simple factors.

Classification by M.G. Valyashko was proposed in 1955. It is based on the same principle of the salt composition. In the degree of detail it practically does not differ from V.A. Sulin's classification. The differences are only in a slightly different systematization and names of the types. For this reason the water types after M.G. Valyashko is convenient to determine using V.A. Sulin's technique but to name them differently. According to the classification by M.G. Valyashko, all natural waters are subdivided by their composition into three basic types: hydrocarbonate (carbonate), sulfate and chloride. To the carbonate type belong waters, in which the HCO_3^- content exceeds the combined content of Ca^{2+} and Mg^{2+}. This type is identical with the 1 class waters after Palmer and with hydrocarbonate–sodium type after V.A. Sulin. The chloride type after M.G. Valyashko includes waters, in which the Ca^{2+} content exceeds the combined content of HCO_3^- and SO_4^{2-} ions. This type is similar to chloride–calcium type after V.A. Sulin. The remaining natural waters are attributed to the sulfate type, which includes two subtypes: sodium and magnesium. The sodium subtype corresponds with the sulfate–sodium and the magnesium, with the chloride–magnesium type after V.A. Sulin.

O.A. ***Alekin's classification*** was proposed in 1948 for systematization of all ground waters and is an attempt to merge together the principles of the domination and salt–formation.

O.A. Alekin subdivided all waters by the dominating anion into classes: hydrocarbonate, sulfate and chloride waters. Within each class by the dominating cation he identified groups of calcium, magnesium and sodium waters. Thus, by the dominating macro–components this classification contains 9 basic ground water groups. Within each group, based on the salt composition, he discriminated the types. O.A. Alekin distinguished 4 basic water types, but each group contains only 3 types (Figure 3.7).

O.A. Alekin attributed to type I waters in which the equivalent content of HCO_3^- is greater than the combined content of Ca^{2+} and Mg^{2+} (carbonate alkalinity is greater than the carbonate hardness). In essence this is a water type without irremovable hardness. It completely coincides with 1st class after Ch. Palmer, hydrocarbonate–sodium type after V.A. Sulin and hydrocarbonate after M.G. Valyashko. Type II after O.A. Alekin includes waters, in which the combined content of Ca^{2+} and Mg^{2+} is greater than the HCO_3^- content but smaller than the HCO_3^- and SO_4^{2-} contents together. This water type coincides with sulfate–sodium after V.A. Sulin and the sodium subtype of the sulfate waters after M.G. Valyashko. The ground water type III, according to O.A. Alekin, joins the waters, in which the combined content of Ca^{2+} and Mg^{2+} is greater than HCO_3^- and SO_4^{2-} contents together. These are waters, in which Ca^{2+} and Mg^{2+} chlorides may form. Actually, type III after O.A. Alekin merges the chloride–magnesium and chloride–calcium

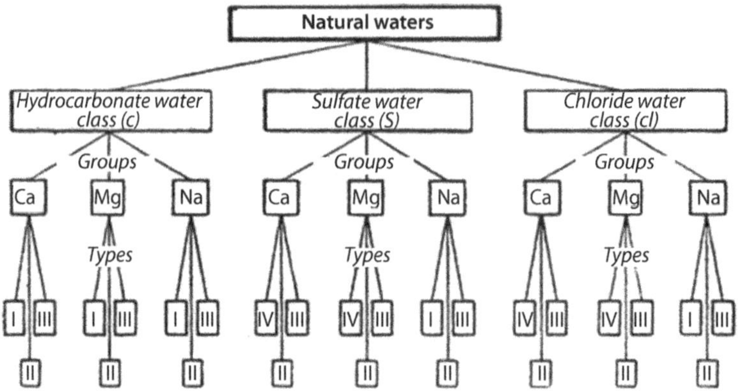

Figure 3.7 Chemical classification of ground waters (after O.A. Alekin).

type waters after V.A. Sulin or the magnesium subtype and chloride type after M.G. Valyashko. At last, type IV after O.A. Alekin includes the waters, in which HCO_3^- is not found. Such acidic water does not have an analog in the classifications after V.A. Sulin and M.G. Valyashko, but in its properties is identical to the waters of Ch. Palmer's fifth class. Thus, the ground water typization after O.A. Alekin in its salt composition, degree of detail and types is closer to the classification after Ch. Palmer.

In substance, all reviewed classifications are based on the same principle, which allows identifying up to five basic ground water types (Table 3.8). Each of these types reflects determined salt properties and also ground water formation conditions.

To the first type are attributed waters in which Na^+ is so plentiful that it is sufficient for Cl^-, SO_4^{2-} and HCO_3^-. For this reason they are called hydrocarbonate-sodium or soda. These are mostly low-salinity soft alkaline waters, in which all hardness is removable. Such waters form in well-flushed sedimentary or crystalline rocks.

To the second type are attributed waters, in which Na^+ is sufficient only for Cl^- and SO_4^{2-}. These waters are called sulfate-sodium and are common in sedimentary rocks and on the surface. Typical for them is relatively low salinity and the presence of irremovable hardness.

The third type waters include those, in which Cl^- is sufficient only for Na^+ and Mg^{2+}. Such chloride-magnesium brines have the most common distribution as they are sea and ocean waters.

To the fourth type are attributed natural waters, in which Cl^- is sufficient for Na^+, Mg^{2+} and Ca^{2+}. These types of waters are called chloride-calcium. These are underground brines with high hardness and unclear formation conditions.

Table 3.8 Comparing ground water chemical classifications after M.G. Valyashko, V.A. Sulin and O.A. Alekin.

	V.A. Sulin, 1946	M.G. Valyashko, 1955	O.A. Alekin, 1948
I	Hydrocarbonate–sodium	Hydrocarbonate	Type I
II	Sulfate–sodium	Sulfate, sodium subtype	Type II
III	Chloride–magnesium	Sulfate, magnesium subtype	Type III
IV	Chloride–calcium	Chloride	
V	–	–	Type IV

At last, it is possible to include in the fifth type acidic ground water, in which there is practically no HCO_3^- but relatively high qualities of heavy metals may be present. These acidic waters with very high CO_2 content are associated with ore deposits and swamps. That is why they are sometimes called carbon–acidic.

N.S. Kurakov proposed to call the natural transition of one type of water to another due to irreversible geochemical processes ***water metamorphism***. The metamorphism direction depends on the nature of the source waters. N.S. Kurakov and M.G. Valyashko consider the water of salt lakes to be sourced by onland fresh water. For this reason they considered the transmutation of low–salinity carbonate–sodium type waters into high–salinity sulfate waters and further to be the ***direct water metamorphization*** and the desalinization as the ***reverse (return) metamorphization*** (Figure 3.8).

If at burial the chloride–magnesium type of sea water converts into higher salinity chloride–calcium type, it is believed that direct metamorphism has occurred. On the contrary, if saline chloride water turns, due to desalinization, into the sulfate–sodium, then into hydrocarbonate–sodium, this process is referred to as reverse metamorphism. For this reason the concentration ratio of Na^+ to Cl^- in equivalent format is often considered as the ground water ***metamorphism factor*** (***coefficient of metamorphism***). In the sea water the value of this factor is equal to 0.87. In a case of direct ground water metamorphization, the value of this factor declines, and in a case of reverse metamorphization it grows.

3.3.2 Graphic Imaging of the Water Composition

Neither Kurlov's formula nor the hydrochemical classifications listed above make it possible to compare natural waters visually. The first graphic

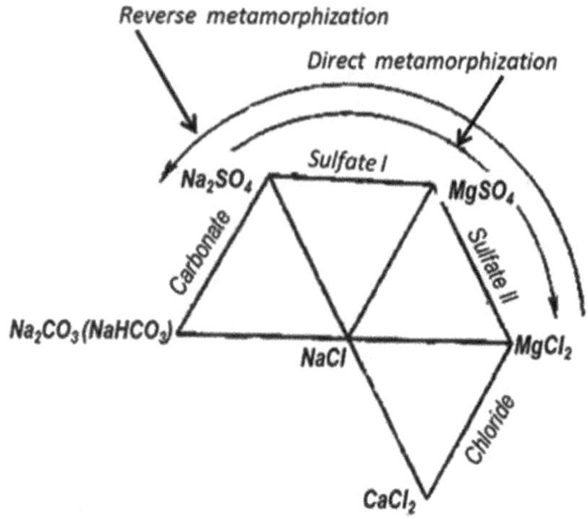

Figure 3.8 Chemical composition of basic salt water's hydrochemical types under M.G. Valyashko classification, and their interrelations in the process of metamorphization.

expression of the water composition was apparently the Palmer–Rogers diagram, proposed as early as in 1917 by G.S. Rogers. Later many other forms were proposed but most commonly used are cyclograms (pie-charts) by N.I. Tolstikhin and X.A. Stiff's patterns.

The ***Palmer–Rogers diagram*** is closely associated with Palmer's classification. It is a graphic image of the water composition in relatively equivalent format (meq-%) presented as two separate columns: anions on the left and cations on the right (Figure 3.9). The columns' height is either 50 meq-% or 100 meq-%. Ion content in the columns is placed down-to-up in descending order of their acidity or alkalinity. These diagrams are a good way for the determination of the water type according to its salt composition and hydrochemical classifications. But it is too unwieldy and inconvenient to use.

The cyclograms (*pie charts*) are a circle divided into sectors. Cations and anions either take different halves of the circle or are positioned at a different distance from its center (Figure 3.10). The circle radius to the assigned scale reflects the value of total salinity. The sectors describe ion content in the relative equivalent format (meq-%). In Russia cyclograms (pie–charts) were introduced by N.I. Tolstikhin in 1933, and they are sometimes called ***Tolstikhin's pie charts***.

In the USA Stiff–patterns developed by H.A. Stiff in 1951 are commonly used. Ion content is presented by a polygon (Figure 3.11) whose left side describes cation contents and the right side, anions. Distance from the central axis is proportionate to individual ion concentrations in equivalent

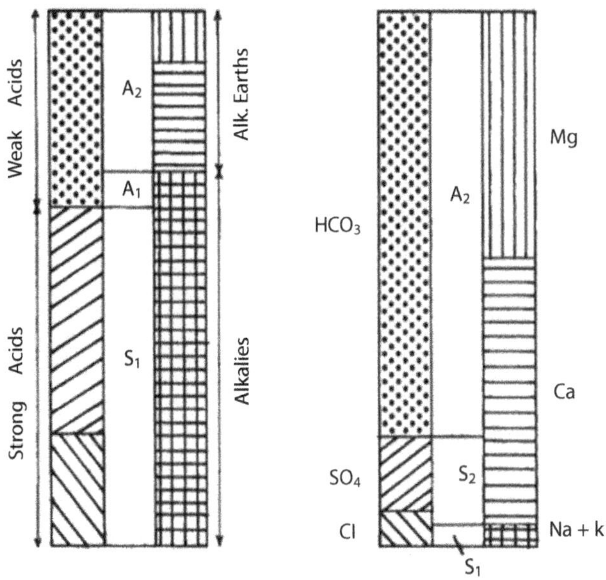

Figure 3.9 Simplified Rogers diagram, in which A and S are alkalinity and salinity after Ch. Palmer, respectively.

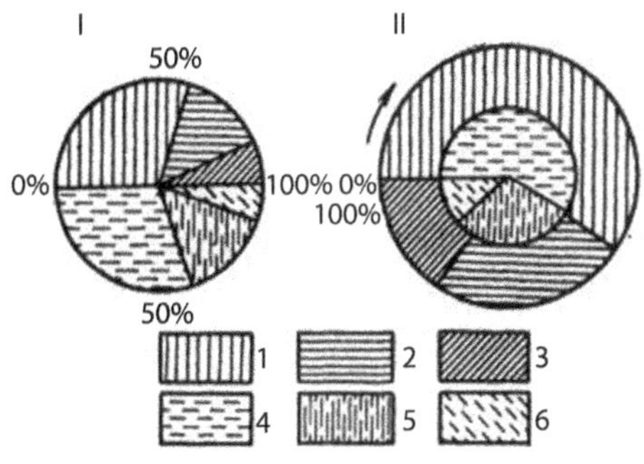

Figure 3.10 Cyclograms (pie-charts). Ions: 1-Ca^{2+}; 2- Mg^{2+}; 3-Na^{+}+K^{+}; 4 -$\boldsymbol{HCO^{-}_{3}}$; 5 - SO_4^{2-}; 6 - Cl^{-}.

format ($meq{\cdot}l^{-1}$). The scale may change from component to component. Usually the scale for Na^+ and Cl^- is 10 times that for the other ions. The vertical positioning of individual ions is not always consistent (compare Figures. 3.11 and 3.12). The perimeter connecting the concentration segments forms a natural pattern whose configuration reflects the composition, the size, and the water salinity. The advantage of this pattern is that compositionally–similar waters have similarly–shaped patterns. Stiff patterns are commonly used for showing water composition on maps (Figure 3.12).

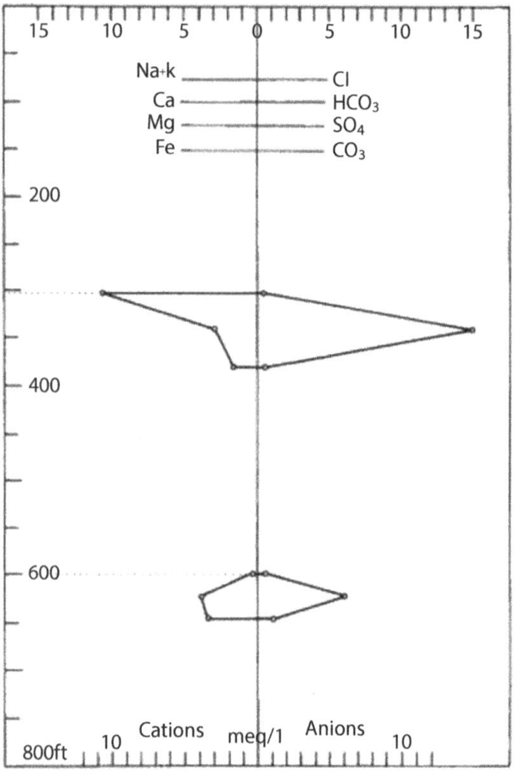

Figure 3.11 Stiff–pattern in mg–equivalents per liter.

The reviewed graphic formats enable visual comparison of ground waters by the content of macro–components, which is very convenient in hydrochemical mapping.

3.3.3 Graphic Comparison of Different Composition Waters

For visually comparing ground water composition graphic methods are commonly used. Initially two–dimensional diagrams were plotted, in which the composition was imaged by a point. The horizontal axis belonged to cations (Na^{+}+K^{+} and Ca^{2+}+Mg^{2+}+Fe^{3+}), and the vertical to anions (Cl^{-}+SO_4^{2-} and HCO_3^{-}+CO_3^{2-}). A drawback of such a technique was in the need to join ions into groups. That is why 3D graphs based on Ferret's triangle were soon introduced. John D. Hem (1916–1994) believed that the first to introduce this technique in 1913 were W.H. Emmons and G.L. Harrington. They used two Ferret's triangles, one for anions, another one for cations. Each vertex corresponded with 100 equivalent–% of Cl^{-},

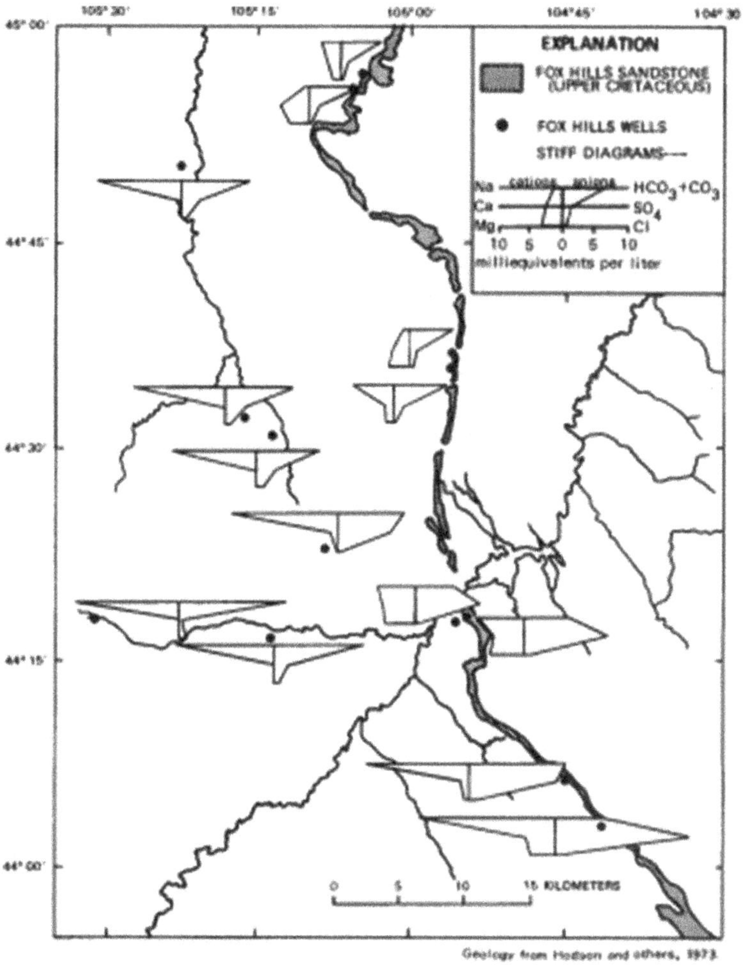

Figure 3.12 Stiff–patterns of macro component concentrations in waters of Fox Hill sandstones, basin Powder River, Wayoming. (Thomas Henderson, 1984).

SO_4^{2-}, HCO_3^-+CO_3^{2-} and Na^++K^+, Ca^{2+}, Mg^{2+}. In this case the water composition is imaged by two points in different diagrams. That was inconvenient for comparing different waters. In 1940 R.A. Hill proposed a technique whereby the points of cation and anion composition in Ferret's triangles were joined in a single field of a 2D diagram (Figure 3.13a). However, most popular in the west became a similar diagram proposed by A.M. Piper in 1944 (Figure 3.13b). In Russia a similar diagram was proposed in 1948 by S.A. Durov (Figure 3.14a). However, it became commonly known only abroad. In Russia much more common was ***Tolstikhin's diagram*** (Figure 3.14b).

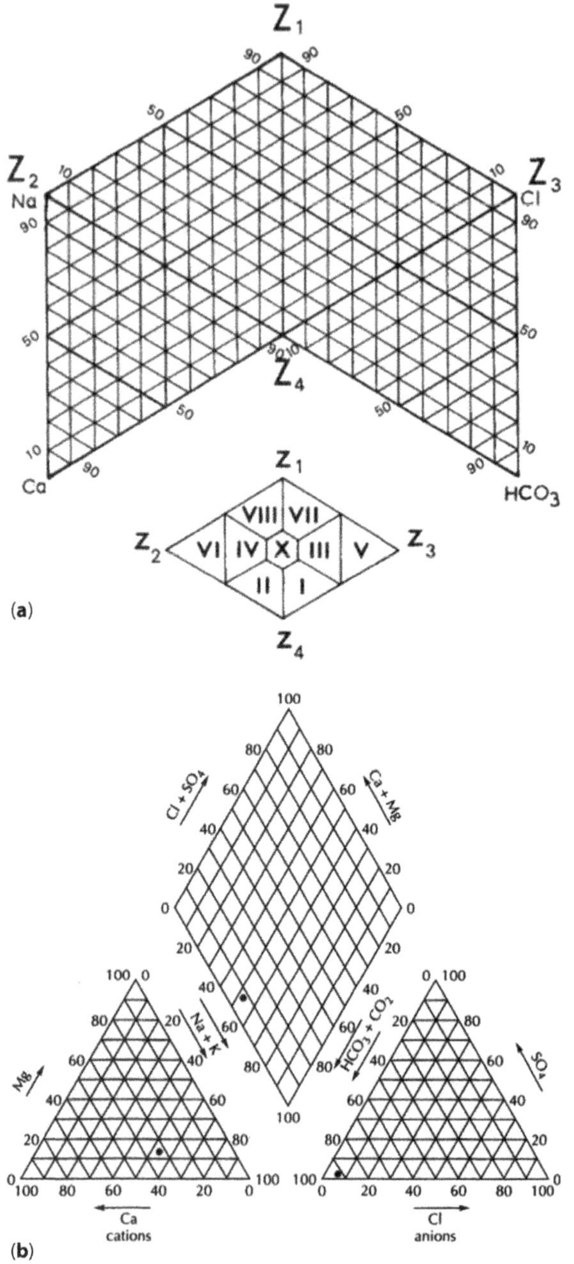

Figure 3.13 Comparison diagrams of waters of different composition: a. R.A. Hill and b. A.M. Piper.

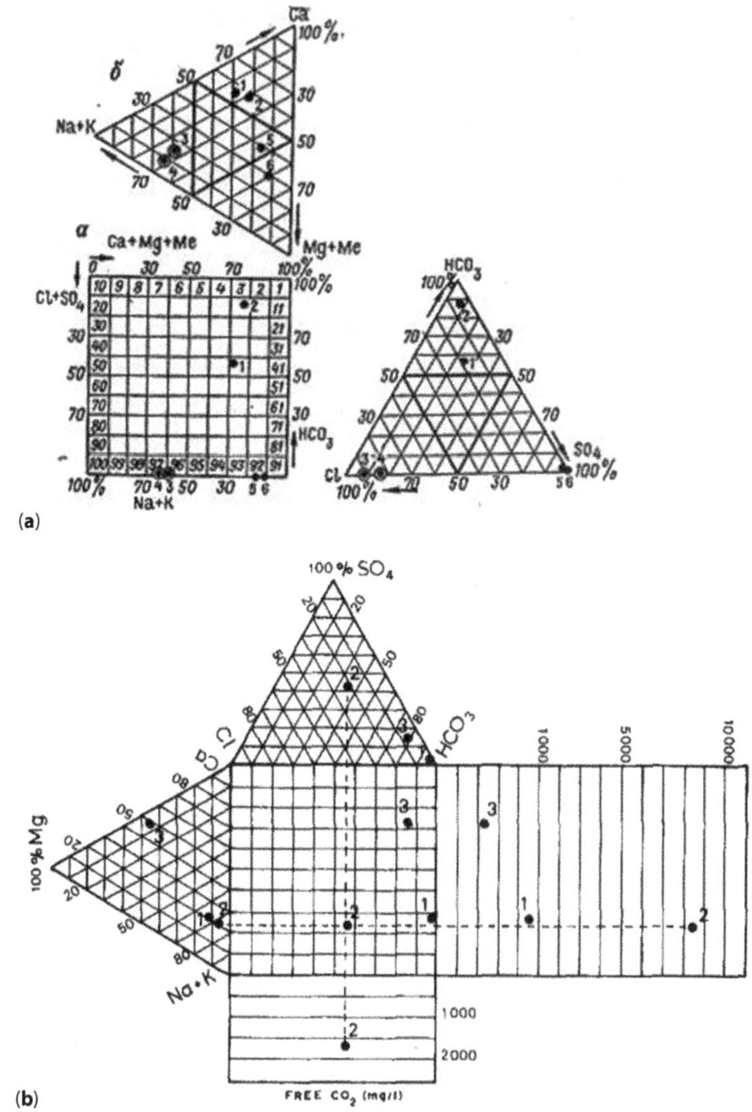

Figure 3.14 Diagrams of comparing waters of different composition: a. N.I. Tolstikhin, b. S.A. Durov.

All aforementioned diagrams were based on the same concept and were very similar to each other. They were a combination of 2 associated Ferret's triangles with a square (S.A. Durov and N.I. Tolstikhin) and rhomb (R.A. Hill and A.M. Piper). Each composition in these diagrams was described by 3 points. The points in the Ferret's triangles determined the ratio of cation and anion concentrations in a relatively equivalent format (meq-%) on the condition that the sums of cations and anions separately were equal to 100%. The final position of the point total composition of water is defined as the intersection of the projections of the points of cations and anions in the square, or diamond.

According to these diagrams, points of waters with similar composition are located next to one another and sequential change in water composition occurs as a trend, i.e., offset of the points in certain direction. More often than others A.M. Paimer's diagram is used.

Plotting of these diagrams takes time and skill. For this reason many program packages intended for hydrochemists also include plotting of these diagrams. In particular, all English-language program packages (*AqQA*, *AquaChem*, *Geochemist's Workbench*, *GW_Chart* and *RockWorks*2006) contain execution of Stiff and Piper diagrams, some of them include pie-charts and Durov diagrams. A simple and free of charge program for plotting the Piper's diagram may be found on the site http://*water.usgs.gov/nrp/gwsoftware/GW_Chart*.

3.3.4 Hydrogeochemical Maps and Cross-sections

Hydrochemical maps and cross-sections are a visualization format of a 3D model. The maps graphically present the idea of regional changes in ground water properties and composition against the background of geographical, geologic or social environment. Hydrochemical cross-sections are a graphical imaging of ground water properties and composition changes with depth against the background of geologic medium. Maps and cross-sections supplement each other.

By the scale and purpose hydrogeochemical maps are usually subdivided into small-scale location maps, medium-scale work maps, and large-scale local and detailed maps.

The small-scale maps are 1:500,000, 1:1,000,000 and smaller. They are applied for the reconnaissance of hydrogeochemical conditions over large areas during the estimation of their potential in terms of fresh, mineral, and industrial ground waters or oil and gas occurrences. For hydrogeochemical tasks small-scale hydrogeochemical maps of a synoptic nature are commonly used. The existing maps of this nature were made based on materials of hydrogeological surveys conducted by geologic directorates, institutes and various ministries of the former USSR. These maps include: a hydrochemical map of the USSR (1956; 1:5,000,000), I.K. Zaytsev, Editor; and a hydrogeochemical map of the USSR (1964; 1:2,500,000), I.K. Zaytsev and N.I. Tolstikhin, Editors. It is also possible to include in this group a specialized map of the USSR mineral waters (1:7,500,000; I960' compiled by V.V. Ivanov, A.M. Ovchinnikov and L.A. Yarotsky) and of the USSR balneological waters (1:4,000,000; V.V. Ivanov, Editor, 1968). All these maps were made by individually developed unique techniques based on working middle-scale maps.

Work maps have scale 1:200,000, 1:100,000 and are usually made based on materials collected directly in the state hydrogeological survey of the same scale. Usually the information about ground water chemical composition is shown in the corresponding hydrogeological maps by color, shading and non-scale letter-digit parameters directly at the points of hydrochemical sampling (wells, springs and shaft wells). Sometimes additional hydrogeochemical maps are constructed covering the entire studied area; depending on the availability of factual material such maps may be certified or uncertified. Middle-scale hydrogeochemical maps are used in the search of ground water and deposits of solid economic minerals as well as for ecological geology studies. They are noticeably different in their information content and in methods of graphic expression.

Local large-scale maps (1:50,000 - 1:25,000), as middle-scale ones, are based on materials of the corresponding state hydrogeological surveys. Hydrogeochemical information in them is also usually represented only in in the form of a description of sampled water points. Sometimes, in case of need and on special order, specialized hydrogeochemical maps are made. They find application for the substantiation of master plans of a territory development in the appraisal and development of individual ground water deposits and in monitoring of contamination from technogenic sources. For making such maps instrumental ties of testing points to the landscape are sometimes used.

At last, detailed hydrogeochemical mapping to a scale 1:10,000 and larger is conducted (and not in all cases) in engineering surveys for specific construction and in the estimation of the scale and evolution of ground water contamination of a great danger.

There is no generally accepted technique of hydrogeochemical mapping. The information load of hydrogeochemical maps is determined by their objective. Hydrogeochemical maps may show general natural properties and composition of ground water in order to estimate their suitability as an economic deposit (drinking, technical, mineral water, etc.); they may be used for the search of economic deposits (ore, oil and gas fields, etc.); and they may describe the ecological state of the subsurface, etc. In any of these applications there are mapped parameters of the general hydrogeological environment and special parameters of ground water composition. Parameters of the general environment are attributed to topography, hydrography, lithology and hydrogeology (discharge areas, hydroisobaths, flow lines, etc.). Parameters of water composition depending on the task may give the idea of the distribution of the salinity, general salt composition, heavy metals, organic and inorganic pollutants, search attributes for the fields of solid economic deposits, etc.

Two basic ways to graphically express the distribution of different quality ground water may be identified: 1. qualitative, by way of zoning a territory, and 2. quantitative, by directly mapping the measurement result values.

Hydrogeochemical zoning is based on identification of the distribution areas of relatively uniform ground water and determination of their separating boundaries. These areas, as a regular rule, are identified by a complex of parameters and are called ***hydrochemical zones.*** Sometimes such zones are identified by genetic attributes. Hydrogeochemical zoning is used mostly in small-scale synoptic mapping and for the description of ground water distribution by general parameters (salt composition, hydrochemical type, etc.)

In the quantitative mapping the major zoning criterion is the value of a specific parameter (the concentrations, their ratios, salinity, etc.). As boundaries are used iso-contours, i.e., contours of equal value of the mapped parameters. Salinity iso-contours are sometimes called ***isomins*** and concentration iso-contours of individual components, ***isochores***. For this reason at the basis of quantitative mapping are methods of interpolation, extrapolation and approximation.

3.3.4.1 Making Hydrogeochemical Maps

The mapping technique radically changed over the recent 20 years. Hand mapping methods on paper were replaced by programmed methods on computer screen. As a result, a new geomapping strategy was formed, under which the information is stored separately as a database on computer and the map image is presented by the computer monitor, printer or plotter. Digital technology allows updating, supplementing and at any moment displaying this information onscreen in different formats including as 2D or 3D geoinformation models. At the basis of these models is well organized and managed information about any landscape objects (topography, geography, hydrography, geology, urbanistics, etc.) in unified space coordinates. Such a system of managed information forms ***the***

geographical information system* - *GIS*.** Its nature is determined by the bases of positional (geographic) and descriptive (attributive) data and programs, which allows managing them. The basis of programs is the ***automated cartographic system (ACS), which is also used for gathering, safekeeping, analysis and graphic visualization of geologic data. Numerous cartographic program packages currently exist based on GIS. The uncontested leader in creating local databases is *ESRI* (*Environmental Systems Research Institute, Inc.*, USA) with program packages *ArcInfo, ArcView GIS, ArcSDE, ArcIMS, MapObjects* and *ArcGIS.* Most common in Russia are program products *ArcGIS* by *ESRI, GeoMedia* of Intergraph Corporation, *MapInfo Professional* by *Pitney Bowes MapInfo* and GIS Map 2008 by ZAO KB "Panorama" and *Isoline GIS.*

With the arrival of GIS and ACS the idea of a digital map was formed, which serves as the basis of their information support. Under GOST 28441—99, ***digital maps*** are digital cartographic models reflecting the content of certain type and scale maps. The content of these maps is perceived due to visualization in the form of ***electronic maps*** imaged on video monitors or ***computer maps*** printed-out as hard copies. Digital map data may be used directly as the source of information for calculations without visualization and transported as databases on memory carriers. Electronic maps give a visual image of the content of digital maps and may also be transported as graphic files (*.gif*, *.jpeg*, *.tiff*, *.bmp*, *.mif*, *.e*00, etc.). At last, computer maps reflect the content of digital maps on hard carriers (paper, film, tablets, etc.). The nature of all these maps obviously depends on the content of used positional and attributive data.

Currently the results of all works for geologic and resource mapping, including the preparation and publication of corresponding geologic map sets, which are conducted by organizations of the RF Ministry of Natural Resources and its contractors (universities and institutes of the Russian Academy of Sciences), are supplied to clients in GIS-format together with numerous digital databases. Requirements to geologic maps are defined under GOST 50836-95 "Geologic cartography. Legends on map of geologic content. General imaging rules". Requirements to digital maps are regulated under GOST 51605-2000 "Maps digital topographic. General requirements" and 28441-99 "Cartography digital. Terminology and definitions".

Making of any hydrochemical map begins with the determination of its purpose and information load, i.e., with constructing a map legend and its symbols.

The symbols may be areal, linear, non-scale and letter-digital. Each type of symbol is entered on the map legend. Areal symbols are used for the description of 2D objects. They correspond with the map scale and are applied in color, shading or speckling. Linear symbols are used for the description of the position of boundaries and other linearly oriented objects (faults, cross-section positions, hydrochemical anomalies, isochores, etc.). These symbols are expressed in lines of various configuration, thickness and color. Non-scale symbols are usually used as points for the description of positions of very small objects (springs, shaft wells, wells, etc.). They are symbols reflecting the essence of the object and may have various size and color. Each non-scale symbol relates to the position of some point on the map; the point is positioned either in the center of the symbol or in one of its corners. Letter and digit symbols describe specific parameters of objects on the map (sequential number, stratigraphy, depths of occurrence, contour digitization, component concentration, etc.). Nondimensional and letter-digit symbols are always oriented along the lower edge of maps.

The symbols may be topographic (geographic), geologic, hydrogeological and special hydrochemical. The topographic parameters are necessary for orienting on the map and for the description of general geomorphic and hydrographic conditions. They include settlements, topographic parameters, rivers, lakes, water points, etc. Geologic parameters are necessary for the description of stratigraphy, lithology, and occurrence conditions of the studied water horizons. General use of imaging and use of symbols on the maps of geologic content to a scale 1:25,000 and smaller and also the cross-sections and columns, regardless of their objective, are defined under GOST P 50836-95 – "Geologic cartography. Symbols on maps of geologic content. General rules of imaging". Hydrogeological parameters are necessary for the description of the distribution, conditions of occurrence and flow dynamics of the studied ground water. At last, special parameters are intended for the description of qualities and composition of the studied waters (salinity, salt type, hardness, content of individual components, functional interrelation of components, etc.).

The summary of all symbols and explanations disclosing their content determines the content of a map legend.

After the information content of the map and its legend are determined the making of the map itself begins. It includes:

- Preparation of the ***base map*** with necessary topographic, geologic and hydrogeological information;
- Plotting on the map testing points with the description of ground water occurrence conditions, their properties and composition;
- Analysis of the spatial distribution of the studied ground water properties and composition parameters.

1. For constructing the base are usually used topographic, geologic and hydrogeological maps to appropriate scale, from which extraneous information is removed. Any maps can be used for constructing such base. In constructing digital maps they are imported as files of various graphical formats: *AutoCAD* [*.dxf*], *Windows Metafile* [*.wmf*], *BitmapGraphics* [*.tif*], [*.bmp*], [*.pcx*], [*.gif*], [*.jpg*] and some others.

In 1994 at the RF Ministry of Natural Resources (currently Rosnedra) the State bank of digital geologic information (GBZGI) was set up and in 2003, the bank of digital cartographic information (BZKI) at the Russian Federal Geologic File. BZKI is an integrated distributed medium for safekeeping of digital geologic–content cartographic information in the territory of Russia intended specially for providing subsurface users with digital cartographic information. BZKI accumulates and integrates information whose base is geologic–content digital map models on the following subjects:

- vector general geographic maps (scales 1:8,000,000, 1:4,000,000, 1:2,500,000);
- unified digital topographic base of geologic–content maps (scales 1:1,000,000, 1:500,000, 1:200,000);
- digital models of geologic–content maps (scales 1:2,500,000, 1:200,000);
- digital geologic atlases and digital integrated geologic information packages.

Most 1:200,000 maps are suitable for the use in programs *ArcInfo*, *ArcView GIS v.3.**, *ArcGIS v.8*, *v.9*.

The base map for electronic construction may be multilayered. If so, each information layer belongs to a separate environment parameter (topography, hydrography, water horizon parameters, hydrodynamic parameters, contamination foci, etc.). At hand–mapping on paper these layers are usually merged in one base map.

2. Sampling points are plotted by hand or automatically in accordance with their coordinates. These points are marked by non–scale symbols, which describe the type of water–show (wells, shaft well, spring, swamp, etc.). Around each water point are written its number and measurement results of its most important parameters (depths to water, salinity, etc.). Often graphic methods of their expression such as pie–charts or Stiff diagrams (Figure 3.12) are used for a description of the water composition. Locations of sampling points with descriptions of their water shows form the ***map of source data*** (***post map***) or map of factual material.

3. Spatial distribution analysis, as a rule, is conducted for qualitative (zoning) or quantitative (contours of equal value) mapping. It is usually performed separately for each parameter.

At qualitative mapping, i.e., zoning, such parameters, as a rule, are complex, often non-numerical descriptions of water properties or composition (water types under different classifications including salt composition, Tolstikhin's pie-charts, stiff patterns or other symbols). Then territories with equal ground water parameters are merged in hydrochemical zones. Boundaries between the fields with different parameters are drawn usually by hand in consideration of all factors, which may affect their position (tectonics, reservoir lithology, topography, flow (filtration) direction, settlements, location and configuration of potential contamination sources, etc.). Examples of qualitative mapping may be hydrogeochemical maps in Figures 3.15 and 3.16, in which zones of waters of different qualities and type are shown, respectively.

Quantitative mapping is based on making contour maps using interpolation and extrapolation techniques. These methods in hydrogeochemistry

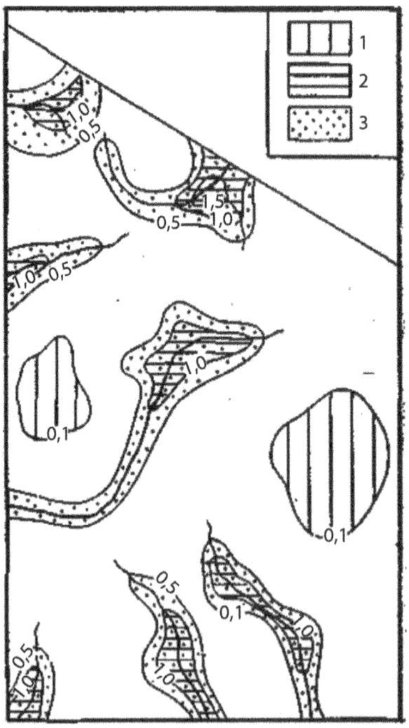

Figure 3.15 Hydrogeochemical map of ground water quality regional estimate: 1. water with hydrocarbonate corrosiveness; 2. water with sulfate corrosiveness; 3. hard water (***Practice for hydrogeochemistry***, Pityeva, K.E. et al., MGU Publishers, 1984).

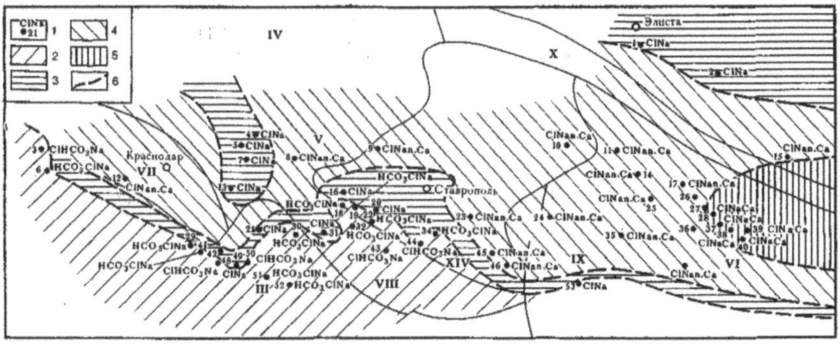

Figure 3.16 Hydrogeochemical schematic map of waterhead complex K_1h–ap in the Fore-Caucasus: 1. No. of water point and chemical composition formula by the component precedence; zones of water distribution with composition: 2. HCO_3–Cl–Na, Cl-HCO_3–Na; 3. Cl–Na; 4 - Cl–Na–Ca; 5– Cl–Na–Ca; 6. Zone boundary (***Practice for hydrogeochemistry***, Pityeva, K.E. et al., MGU Publishers, 1984).

have limitations for parameter values and for interpolation distance, which determine contour intervals.

A first limitation is associated with the necessity to map only significative differences between measurement results; that is especially important with short distances between sampling points (when Δx is close to 0). In variograms this limitation is associated with the presence of the nugget effect, namely, with a large standard deviation value of global error σ_i. Criteria of significative differences between two mean measurement results have been shown earlier. They depend on global errors ΔC_i as well as on a number of repeated testing events n. In particular, at confidence level 0.95 the interval between values of the adjacent contours must be greater than:

$$\text{at}\, N = 1,\ \ 2.77\sigma_i = 1.41\Delta C_i;$$
$$\text{at}\, N = 2,\ \ 1.96\sigma_i = \Delta C_i;$$
$$\text{at}\, N = 3,\ \ 1.60\sigma_i = 0.82\Delta C_i.$$

If these testing technique parameters are absent and a standard selection's deviation is used, then at confidence level 0.95 intervals between adjacent contours must exceed:

at $N = 2$, $4.3s_i$,

at $N = 3$, $2.3s_i$,

at $N = 4$, $1.7s_i$,

where s_i is a selection's mean standard deviation from n repeated testing events.

A second limitation is related to the maximum distance between sampling points at which the interpolation is acceptable. According to variograms, parameter values at two sampling points are functionally associated at a distance less than the range effect. For this reason the interpolation may be considered reliable if the distance between sampled points is less than the values of this range effect.

There are numerous contouring interpolation techniques, primitive hand-drawing to complex computer–driven. The simplest hand technique assumes a linear correlation of variable parameter value vs. distance between testing points. Testing points are connected by straight lines, the result is a number of triangles. Then on these straight lines by way of interpolation are posted points where the values of a variable parameter are equal to the contour values. This technique is relatively simple and flexible, and it allows taking into account locations of provenance areas, directions of ground water flow (filtration) and discharge, change of mineral composition in inclosing rocks' waters, etc. However, this technique is good over a limited area at a low number of testing points and obviously deterministic correlation of the parameter value vs. space coordinates. Examples of hydrochemical maps with contours are given in Figures. 3.17 and 3.18. They show the salinity and sodium development in waters of individual complexes in the Fore–Caucasus.

The automatic interpolation is done in two steps. First step: parameter values of randomly–positioned testing points are interpolated into intersection points of a ***regular grid***. Most popular are orthogonal grids with square or quadrangle cells in cartographic projection. This transitional procedure from a nonuniform source data point distribution to uniform

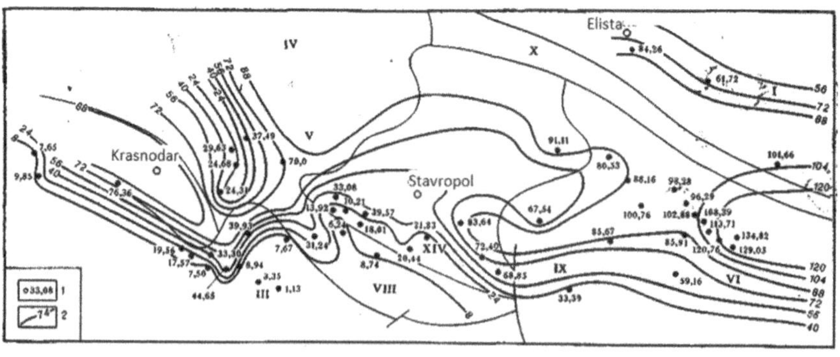

Figure 3.17 Schematic map of head ground water salinity in K_1h – ***ap*** complex of Fore–Caucasus: 1. ground water salinity value, g·l^{-1}, 2. Salinity contours, g·l^{-1}.

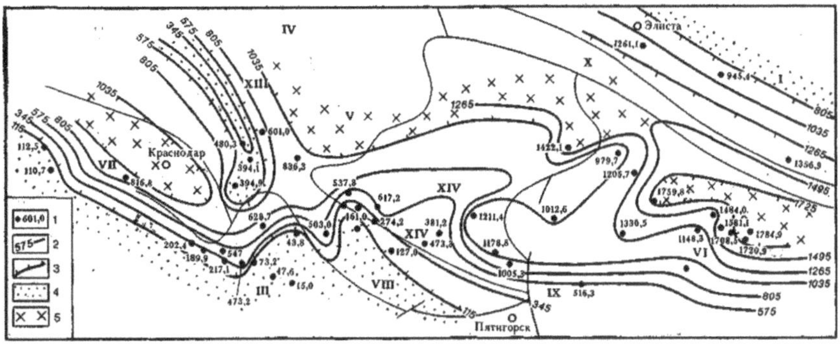

Figure 3.18 Schematic map of sodium distribution (mg–equiv/l) in ground waters of K_1h– ***ap*** complex, Fore–Caucasus (1. testing points with Na concentration values, 2. Na concentration contours, 3. Regional faults, 4. Abnormally–low Na concentrations, 5. Abnormally–high Na concentrations).

distribution, strictly in the ***grid nodes***, is called ***griding***. The result of this procedure is a ***grid map***, in which points with values of a variable parameter are distributed regularly and uniformly. The grid size and interpolation technique are assigned beforehand. Derived values of the measured variable at the points with orderly surface coordinates form a new selection with orderly ***grid data***, which serve a basis for generating regional maps.

The most used program providing for a high quality of 2D contouring and interpolation in conditions of 3D surface mapping was the *Surfer* program package developed by "*Golden Software Inc.*". This package proposes 12 interpolation algorithms, including methods of *inverse distance to a power, kriging, minimum curvature, radial basis functions, polynomial regression,* etc. Most commonly used among the suggested interpolation techniques is geostatistical interpolation technique – kriging.

Kriging includes a whole number of geostatistical interpolation techniques based on computing the sliding averages of variable values weighted by the distance to the adjacent points with established parameters. In the simplest case of ***ordinary kriging*** the sought–for value $C_{i,0}$ of any parameter at unsampled point x_0 is calculated as the weighted average value of the same parameter at the adjacent sampled points from the following equation:

$$C_{i,0} = \sum_{n=1}^{N} \lambda_n \overline{C}_{i,n}$$

where $\overline{C}_{i,n}$ is parameter i measurement result at N points positioned at a distance x_n ($n = 1, 2,\ldots,N$) from the interpolation point, λ_n is weights of these results (Figure 3.19). Their weights are functions of the distance x_n

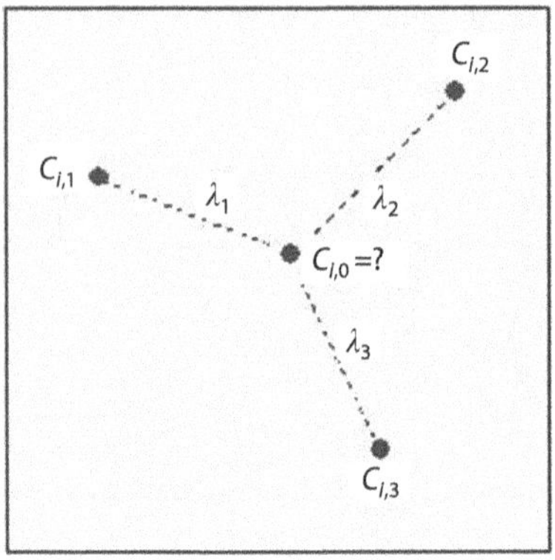

Figure 3.19 Interpolation by average weighing of a parameter value at surrounding points.

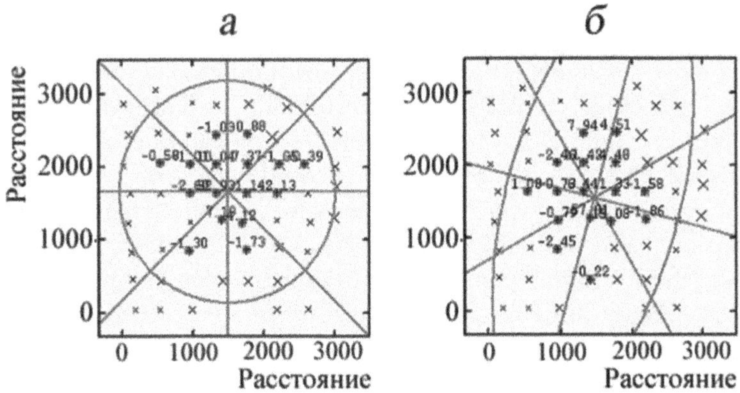

Figure 3.20 Effect areas in isotropic and in anisotropic media.

effect, and the sum of these distance is equal to 1. Weighing factor λ_n plays here the role of regression factor, which is selected so that the sought-for value $C_{i,0}$ would have minimal systematic and random errors.

Usually not all measurement results around point x_0 are used, but only those associated with value $C_{i,0}$ by a functional correlation, i.e., are within the influence area. This influence area is within the circle (isotropic medium) or ellipsis (anisotropic medium) with a radius equal to the variogram range effect (Figure 3.20). Kriging's advantages are in that it is using metamodels of directional variograms, which allows considering their statistical trends, variable's autocorrelation factor association with the distance and the spatial length of this effect. The less the range effect on the variogram, the greater the testing density of any parameter required for reliable mapping. At the

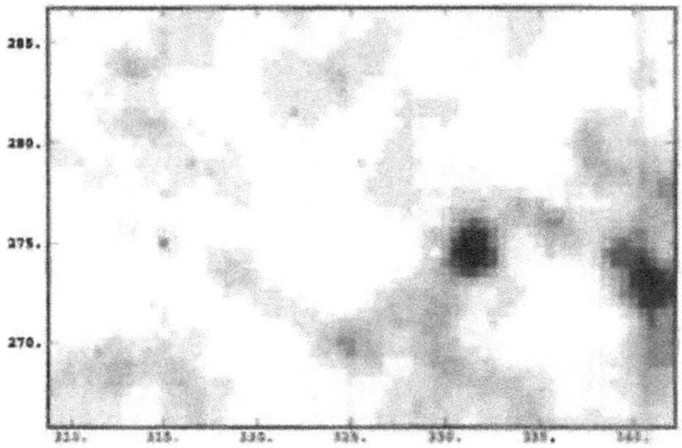

Figure 3.21 Raster visualization of arsenic concentrations calculated by kriging technique.

same testing density the mapping reliability by interpolation and extrapolation decreases together with the decrease of the range effect on variogram. In the case of very short range effect or totally absent range effect (variograms of pure nuggets) the interpolation and extrapolation methods are unacceptable.

Orderly data are a basis for the construction of regional maps. For this purpose their source data are used again for interpolation of the variable values in the nodes of a finer grid. Sequential decrease of grid cells and increase in the number of orderly data give smoother distribution of the parameter value. If after that the parameter values in individual cells of grid–maps are highlighted by color or another symbol in accordance with the selected symbols, the result is a ***raster map*** (Figure 3.21). The separation of areas in these raster maps by lines results in a contour map with contours of equal values of the mapped parameter (Figure 3.22).

All operations are performed automatically according to directions of the operator.

The advantage of the automatic mapping is in its capability to rapidly generate maps–models with different contour intervals using different interpolation methods. The generation of such maps, their comparison and analysis in order to select the optimal option is the essence of ***cartographic simulation***. In this case maps are just different mathematical models of the mapped property with the preservation of spatial statistics (geostatistics) of the source data. And each separate model map is one of many possible estimates of the state of the studied object. A disadvantage of the automatic mapping is its impossibility in most cases to account for the effect on the mapped value of various unquantitative factors, which usually are a priori taken into account by a live researcher (configuration of areal contamination sources,

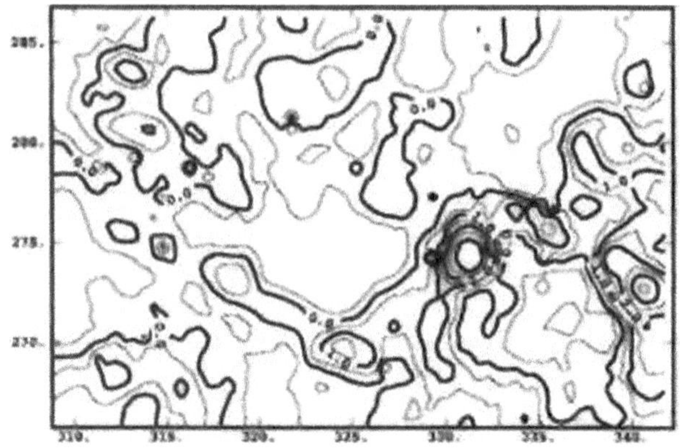

Figure 3.22 Contour map of arsenic concentrations derived by kriging.

lithological boundaries, ground water direction in local areas, etc.). This disadvantage is most significative with a small number of sampling points.

3.3.4.2 Generating Hydrogeochemical Cross-sections

Hydrogeochemical profiles (cross-sections) are intended for the description of changes in underground water composition and properties with depth in certain regional direction (across or along the strike of water horizons, along the flow lines, etc.). The selection of a direction, content and profile construction technique depend on the assigned tasks and a sufficient amount of factual material. Usually they serve as attachments to hydrogeochemical maps but may also have independent significance.

The generation of a cross-section begins with the assignment of its task and selection of its direction and length on the map. After this a hydrochemical cross-section is constructed as follows:

1. Topographic, then geologo-lithological cross-section is selected or constructed to the scale of the hydrogeological map; it will be the basis of the cross-section. If a digital map is available, the cross-section is constructed using the appropriate program package.
2. Plotted on the cross-section are the lines of wells with objects (intervals) of hydrochemical testing and piezometric levels shown.
3. Next to the testing objects are written derived composition, properties or quality parameters (salinity, concentrations, pH, Eh values, water type, etc.).

4. Based on the assigned task, zone intervals or contour intervals are selected for the component shown on the cross-section. Under the quantitative mapping, intervals between contours (isochores) should exceed $2\,\Delta C_i$.
5. Within each water horizon there are either shown equal value contours of the mapped parameters using interpolation technique or identified distribution zones of different quality waters.
6. In a case of several water horizons, sometimes it is acceptable to merge their zones or equal value contours with crossing the separating water barriers.

There are currently numerous special program packages, which include generating of geologic maps and cross-sections, mostly for petroleum geology. Out of a number of foreign programs two that deserve attention are *RockWorks* and *RockWare GIS Link* by *RockWare.* Of interest in the Russian Federation are program packages "Geozor-Razrez" by the Kazan "Nedra" science-technical center and "Baspro Optima" by ZAO "Baspro" group of companies. Examples of hydrochemical cross-sections to different scales and of different purposes are given in Figures. 3.23, 3.24 and 3.25.

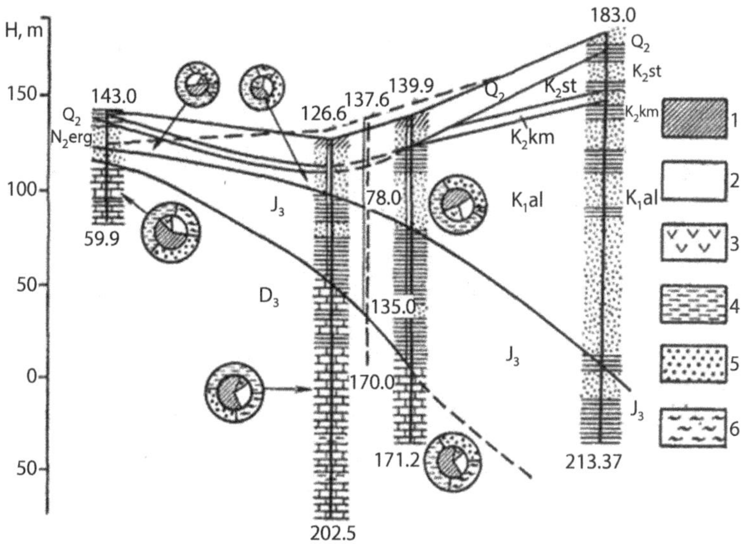

Figure 3.23 Hydrochemical profile through Michurinsk–Tambov– Kirsanov. In Tolstikhin's pie-charts: 1– Ca2+; 2 –Mg2+; 3 – Na++K+; 4 – HCO3–; 5 – SO42–; 6 –Cl–. (Gavich, I.K., Luchsheva, A.A. and Semenova-Yerofeyeva, S.M., 1985).

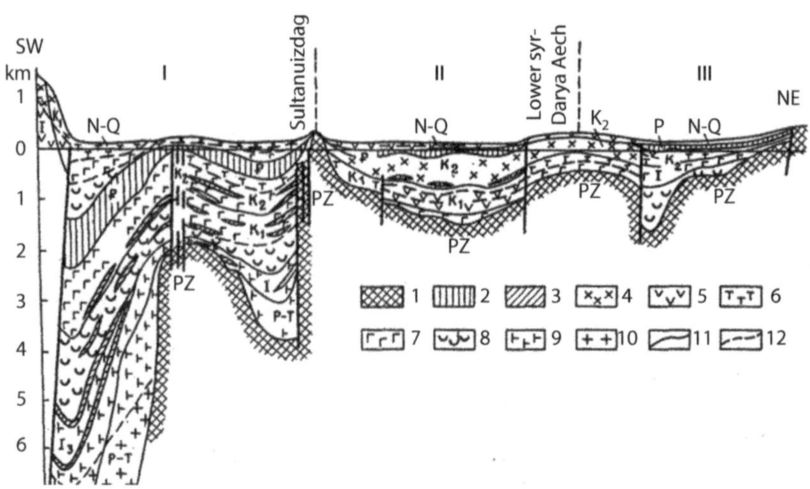

Figure 3.24 Hydrogeochemical cross-section through Amu-Darya (I), Ayr-Darya (II) and Turgay (III) artesian basins. 1. Folded Paleozoic basement; 2. Regional Paleogene water-barrier (seal); 3. Local water-barriers (seals); 4–10. Ground water salinity, g/l: 4. Up to 3; 5. 3–10; 6. 10–35; 7. 35–70; 8. 70–140; 9. 140–270; 10. >270; 11–12. Boundaries (11. Of water complexes, 12. Of hydrochemical zones) (Kraynov, S.R., Shvets, V.M., 1992).

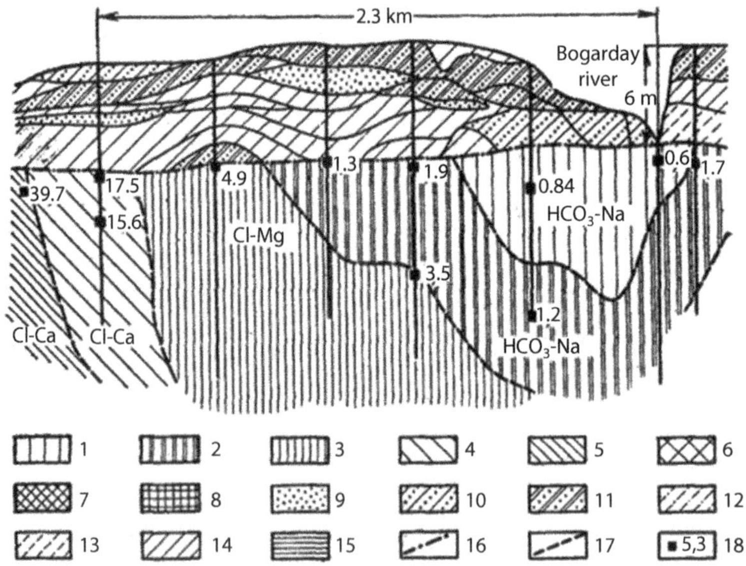

Figure 3.25 Hydrochemical ground water zoning under river channels (North Caspian). Values of water salinity, g/kg: 1. < 1.0; 2. 1–3; 3. 3–10; 4. 10–35; 5. 35–50; 6. 50–75; 7. 75–100; 8. > 100. Rock lithology above the ground water level: 9. Sand; 10. Light sandy loam; 11. Heavy sandy loam; 12. Light loam; 13. Medium loam; 14. Heavy loam; 15. Clays; 16. Ground water level; 17. Boundaries of different water salinity; 18. Depths of water sample collection and water salinity. HCO3—Na, ETC.– water type, Kurakov- Vlyashko classification. (Samarina, V.S., 1977).

Symbols

Symbol	Meaning
Latin	
a	value inverse to range effect of semivariance function.
B_i	Bunsen solubility factor of component *i*.
C_i	relative molar concentration (molar fraction) of component *i*
$[C_i]$	relative ion *i* concentration in equivalent format (meq-%)
$C_{w,i}$	weight concentration of component *i* ($mg \cdot l^{-1}$).
$C_{m,i}$	molality of component *i* ($mole \cdot kg^{-1}$).
$C_{M,i}$	molar concentration (molarity) of component *i* ($mole \cdot L^{-1}$).
$C_{o,i}$	relative volume concentration of component *i* ($ml \cdot l^{-1}$).
$C_{eq,i}$	ion *i* concentration in equivalent format ($mg\text{–}equiv \cdot l^{-1}$).
$\overline{C}_i$	mean value (mathematical expectation) of observed values (measurement result) of parameter *i*.
$\overline{\overline{C}}_i$	mean value (mathematical expectation) of measurement results of parameter $\overline{C}_i$.
$\hat{C}_i$	reference value of parameter C_i.
ΔC_i	parameter C_i observed values absolute global error at one sampling point.
ΔC_i	absolute random error of parameter C_i observed values at one sampling point.

$\hat{C}_i$	reference value error of measuredoro parameter $\hat{C}_i$.
d_i	operative repetitiveness limit (standard) control (convergence) of parameter *i* observed values
D_i	operative reproducibility limit (standard) control of parameter *i* observed values
f	number of degrees of freedom.
F	Faraday's constant (96,484.56 cal/mole or 23.6 Kcall/V·g–eqm).
F_t	Fisher's criterion.
g_i	mass of component *i* in solution.
h	depths to neutral temperature layer.
H	depths to sampling points.
K_i	operative accuracy limit (standard) control of parameter *i* observed values in one testing point.
$\widetilde{K}_i$	distribution factor of component *i* between hydrophobic extractant and water solution.
k	sequential number of lag.
L	number of laboratories.
M_lL_m	salt or complex compound composition (*M* and *L* – cations and anions, *m* and *l their* number, respectively).
M_i	molecular mass of component *i*.
m	salinity (the sum of weighting analytical concentrations of mineral components in water).
N_{col}	number of identified E. coli colonies.
N_i	number of molecules or moles in solution, and number of observed values of parameter i at the same testing point.
$\overline{N}_i$	number of testing points (measurement results) of one parameter *i*.
N_r	number of rectifications at distillation.
p_i	partial pressure of component *i*.
P	probability value of confidence level or interval.
$P_{пл.}$	formation pressure.
P_i^0	saturated steam pressure of component *i*.
Q_g	gas–saturation or gas–water ratio value
$Q(n)$	critical range factor.

Q_t	Cochran's criterion.
R	gas constant (8.314 j·mole^{-1}·K^{-1} or 1.987·10^{-3} Kcall·mole^{-1}·K^{-1})
$\overline{R}$	correlation, autocorrelation or determination factor
$S_{o,i}$	smell threshold number for volume concentrations.
s_i	standard (mean–square) deviation of global error of parameter i in limited selection of observed value at one testing point.
$\underline{S}_i^2$	$\underline{S}_i^2$ - standard (mean–square) deviation of bias of parameter i in limited selection of observed value at one testing point.
$\underline{\underline{S}}_i^2$	$\underline{\underline{S}}_i^2$ - standard (mean–square) deviation of random error of parameter i in limited selection of observed value at one testing point.
$\overline{S}_i$	$\overline{S}_i$ - standard (mean–square) deviation of measurement results of parameter i in selection at one testing point.
T	absolute temperature.
T_C	temperature in degrees Celsius.
T_F	temperature in degrees Fahrenheit.
t_t	Student's criterion.
t	time.
u	quantile.
V_i	volume.
x, y, z	space coordinates.
z_i	ion charge.
Greek:	
α	proportionality factor.
α_i	relative volatility of moisture's component *i* in water steam.
a_i^0	component *i* selectivity factor.
β	proportionality factor.
β_w	waters compressibility factor
$\gamma_{i,k}$	parameter *i* semivariogram's function at sequential lag number *k*
γ_w	water specific gravity
δi	deviation value of measured isotope composition from some standard ratio, in percentages or promiles
δ_i	relative global error of parameter *i* measurement results

δ_i	relative systematic error of parameter i measurement results
$\underline{\underline{\delta}}_i$	relative random error of parameter *i* measurement results
$\bar{\delta}_i$	variation factor of parameter *i* measurement results
Δ	measurement result error (half width of interval where sought–for value is found with accepted probability $P = 0.95$).
$\underline{\underline{\Delta}}_i$	absolute random error or standard error of parameter *i*.
$\pm\Delta_i$	confidence error of parameter *i* measurement results
ε_i	degree of component *i* separation or extraction from the sample.
η_w	dynamic viscosity $Pa \cdot s = N \cdot s \cdot m^{-2} = kg \cdot m^{-1} \cdot s^{-1}$
θ_i	absolute systematic error of parametera *i* measurement
Θ	deviation angle from abscissa (x) coordinate
κ_i	concentrating or dividing factor
λ_i	weights of sampled points relative interpolation point by parameter *i*
ν_w	kinematic viscosity
$\rho_{i,k}$	parameter *i* autocorrelation factor at sequential lag number *k*
ρ_w	specific density in $kg \cdot l^{-1}$
σ_i	standard (mean–square) deviation of general population of parameter *i* observed values at one testing point.
$\bar{\sigma}_i$	standard (mean–square) deviation of general population of parameter *i* measurement results
σ_w	specific electric conductivity, $\mu Sm \cdot cm^{-1}$
ϕ_w	water fluidity.
ω_w	specific electric resistivity

Used abbreviations

BOD	biochemical oxygen consumption
COD	chemical oxygen consumption
urfactant	surface–active matter
DL	detection limit
QL	quantification limit
ACF	autocorrelation function
TOC	total organic carbon
OM	organic matter
Eh	oxidation–reduction potential (ORP) value as measure of **electron activity expressed in** measurement **units of** electric voltage.
pH	hydrogen parameter as hydrogen activity measure in solution, is equal й valueenegativelyго десятичного логарифма this activity.
Pe	oxidation–reduction potential (ORP) value as measure of **electron activity expressed by the** value of negative Brigg's logarithm of their activity.

References

1. Alekin O.A. *Basics of hydrochemistry*. Leningrad, Gidrometeoizdat, 1970, 442 c. (in Russ.)
2. Alley W. Moscow, *Regional Ground-Water Quality*, John Wiley and Sons, 1993. 634 p.
3. Appelo C. A. J., Postma D. *Geochemistry, groundwater and pollution*. Balkema, 2005, 649 p.
4. Barcelona M. J., *Contamination of ground water: prevention, assessment, restoration*, 1990, 213 p.
5. Bodnar, R. J., and J. K. *Costain Effect of varying fluid composition on mass and energy transport in the Earth's crust*, Geophys. Res. Lett., 1991, 18(5), 983-986.
6. Brodsky A.A. *Hydrochemical technique search ore fields (Temporary guidelines)*. Госгеолтехиздат, Moscow, 1957 (in Russ.)
7. Csuros M. *Environmental sampling and analysis for technicians*, Lewis Publishers, 1994, 320 p.
8. Deutsch W. J. *Groundwater geochemistry: fundamentals and applications to contamination*, 1997, 221 p.
9. *Directory enlarged estimated standards for geological exploration*, iss. 1. Gosgeoltehizdat, 1954, 89 p. (in Russ.)
10. Gavich I.K., Semenova S.M., Shvets V.M. *Hydrogeochemical information processing methods with variants of the problem*, Moscow, 1981, 160 p. (in Russ.)
11. Gavrishin A.I. *Evaluation and control qualities geochemical information*. Moscow, 1980, 287 p. (in Russ.)
12. Goroyan V.I., L. A. Grubov, N.M. Kruglikov N.M., Nesmelova Z. N/. *Toolkit for sampling and analysis of natural gas* / - Leningrad, 1969. -160 p. (in Russ.)
13. Henderson T. "Geochemistry of ground-water in two sandstone aquifer systems in the northern Great Plains in parts of Montana, Wyoming, North Dakota, and South Dakota. U.S. Geol. Survey prof. paper"; 1402-C, 1984, 84 p.

14. Katz, D.L., D.Cornell, R. Kobayashi, J.R. Elenbaas, F.H. Poettmann, C.E. Weinaug *Handbook of Natural Gas Engineering*, McGraw-Hill, New York, 1959, 802 p.
15. Kebbekus B. B., Mitra S. *Environmental chemical analysis*, Blackie Academic & Professional, 1998, 330 p.
16. Kirukhin V.A., Korotkov A.I., Shvartsev S.L. *Hydrogeochemistry*. Moscow, Nedra, 1993, 383 p. (in Russ.)
17. Kolotov B.A. *Basics of Hydrogeochemical search ore fields*, 1983 (in Russ.)
18. Kortsenshtein V.N. *Technique of hydrogeological study oil and gas regions*. Moscow,Nedra, 1991, 419 c. (in Russ.)
19. Kraynov S.R., Ryzhenko B.N., Shvets V.M. *Hydrogeochemistry*. Moscow, Nedra, 1992, 463 c. (in Russ.)
20. Lesage S., Jackson R. E., *Groundwater contamination and analysis at hazardous waste sites*, M. Dekker, 1992, 545 p.
21. Loconto P. R. *Trace environmental quantitative analysis: principles, techniques, and applications*, Marcel Dekker, 2001, 646 p.
22. Lukner L., Shestakov V.M. *Modeling migration of ground water*. Moscow, Nedra, 1986, 208 p. (in Russ.)
23. Lurie J.J. *Analytical chemistry of industrial waste water*. Moscow, Chemistry, 1984, 448 p. (in Russ.)
24. Macioszczyk A. *Hydrogeochemia*. WG, Warszawa, 1987, 475 p.(Pol.)
25. Mohamed A.-M. O., Antia H. E., *Geoenvironmental engineering*, 1998, 707 p.
26. Nielsen G., *The essential handbook of ground-water sampling*, CRC Press/ Taylor & Francis, 2006, 309 p.
27. Nollet L. M. L. *Handbook of water analysis*, Marcel Dekker, 2000, 921 p.
28. Otto M. *Analytical Chemistry*. Weinheim, Germany, 2000, p. 453.
29. Perelman A.I. *Hydrochemistry ground water*. Moscow, Science, 1982, 154 p. (in Russ.)
30. Pitieva K.E., Brusilovsky S.A., Vostrikova L.Ju., Chesalov S.M. *Workshop on hydrogeochemistry*, Moscow, Pub. MGU, 1984, 254 c. (in Russ.)
31. Popek E. P. *Sampling and analysis of environmental chemical pollutants*, Academic Press, 2003, 356 p.
32. Reznikov A.A., E.P. Mulikovskaya, I.Y. Sokolov, *Methods analysis ground water*. Moscow, Nedra, 1970, 488 p. (in Russ.)
33. Rogovskaya N.V. *Hydrogeologicaloe mappinge*. Moscow, Science, 1981. 132 p. (in Russ.)
34. Samarin V.S. *Hydrogeochemical testing ground water*, Leningrad, Pub. LGU, 1958, 257 p. (in Russ.)
35. Semenov A.D. *Guidance on chemical analysis of surface waters*, Leningrad, Gidrometeoizdat, 1977.- 541p. (in Russ.)
36. Shvartsev S.L. *Hydrochemical supergene zone*, Moscow, Nedra, 1998, 366 p. (in Russ.)
37. "Site Characterization for Subsurface Remediation. Seminar Publication." // EPA/625/4-91/026, 1991, 259 pp.

38. Touloukian Y. S., Saxena S. C. and Hestermans P., "Thermophysical Properties of Matter. Viscosity," vol. 11, *The TPRC Data Series*, IFI/Plenum, New York, 1975
39. Weight W. D., Sonderegger J. L., *Manual of applied field hydrogeology*, McGraw-Hill Professional, 2001, 608 p.
40. Zhang Ch., *Fundamentals of environmental sampling and analysis*, Wiley-Interscience, 2007, 436 p.

Standard Publications

GOST 13273-88, "Water. Mineral, drinking, balneological and curative-table. Technical conditions."
GOST 18164-72, "Water potable Dry residue content determination technique."
GOST 18963-73, "Water potable. Methods of sanitary-bacteriological analysis"
GOST 23268.0-91, "Water mineral, drinking, balneological, curative-table and natural table. Rules and methods of sample collection."
GOST 23268.12-78, "Water mineral, drinking, balneological, curative-table and natural table. Permanganate oxidability determination technique."
GOST 23268.2-91, "Water mineral, drinking, balneological, curative-table and natural table. Carbon dioxide determination methods."
GOST 24849-81, "Water potable. Field methods of sanitary-micro-biologic analysis".
GOST 24902-81, "Water domestic/potable. General requirements to field analysis methods"
GOST 27065-86, "Water quality. Terminology and definitions."
GOST P 27384-2002, "Water. Errors measurements standards of composition and property parameters."
GOST 2874-82, "Water potable. Hygienic requirements and quality control."
GOST 29131-91, "Products liquid chemical. Color measurement technique in Hazen units (platinum-cobalt scale)."
GOST 3351-74, "Water potable. Taste, smell, chromaticity and turbidity determination methods."
GOST 4151-72, "Water potable. Total hardness determination technique."
GOST 4796-49, "Concrete hydrotechnical. Corrosiveness indications and standards of water-medium. Standardgiz, Moscow, 1949."
GOST P 50836-95, "Geologic cartography. Legends on maps of geologic content. General rules of imaging."
GOST P 51000.4-96, "State standardization system in Russian Federation. Certification system in Russian Federation. General requirements to certification of testing laboratories."
GOST P 51209-98, "Water potable. Determination technique of chlorine-organic pesticides content by gas-liquid chromatography."

GOST P 51211–98, "Water potable. Determination techniques of surface-active substances content."

GOST P 51232–98, "Water potable. General requirements to logistics and methods of quality control."

GOST P 51310–99, "Water potable. Determination technique of benzopiren content."

GOST P 51392–99, "Water potable. Determination of volatile halogen–organic compounds' content by gas–liquid chromatography."

GOST P 51592–2000, "Water. General requirements to sample collection."

GOST P 51593–2000, "Water potable. Sample collection."

GOST **P** 51605–2000, "Maps digital topographic. General requirements" and GOST 28441–99 "Cartography digital. Terminology and definitions."

GOST P 51730–2001, "Water potable. Determination technique of total specific alpha–activity of radio–nucleids."

GOST P 51797–2001, "Water potable. Petroleum products' content determination technique."

GOST P 52029–2003, "Water. Units of hardness."

GOST P 52991–2008, "Water. Total and dissolved organic carbon content determination methods."

GOST P 6709–72, "Distilled water. Technical conditions."

GOST P 8.563–96, "State system of measurement units' unity assurance. Measurement performance techniques."

GOST P ISO 5725–2002, "Accuracy (correctness and precision) of measurement techniques and results."

GOST 9.602–89, "Unified protection system against corrosion and ageing. Underground facilities."

SanPiN 2.1.4.1074–01, "Potable water. Hygienic requirements to water quality in centralized systems of potable water–supply. Quality control."

SNiP 2.03.02–86, "Concrete and reinforced concrete constructions from compact silicate concrete."

Index

Also of Interest

Check out these other titles from Scrivener Publishing

Seismic Loads, by Victor Lyatkher, ISBN 9781118946244. Combining mathematical and physical modeling, the author of this groundbreaking new volume explores the theories and applications of seismic loads and how to mitigate the risks of seismic activity in buildings and other structures.

Hydraulic Modeling, by Victor Lyatkher and Alexander M. Proudovsky, ISBN 9781118946190 Combining mathematical and physical modeling, the authors of this groundbreaking new volume explore the theories and applications of hydraulic modeling, an important field of engineering that affects many industries, including energy, the process industries, manufacturing, and environmental science. *Publishing in January 2016.*

Fundamentals of Biophysics, by Andrey B. Rubin, ISBN 9781118842454. The most up-to-date and thorough textbook on the fundamentals of biophysics, for the student, professor, or engineer. *NOW AVAILABLE!*

i-*Smooth Analysis: Theory and Applications*, by A.V. Kim, ISBN 9781118998366. A totally new direction in mathematics, this revolutionary new study introduces a new class of invariant derivatives of functions and establishes relations with other derivatives, such as the Sobolev generalized derivative and the generalized derivative of the distribution theory. *NOW AVAILABLE!*

Reverse Osmosis: Design, Processes, and Applications for Engineers 2nd Edition, by Jane Kucera, ISBN 9781118639740. This is the most comprehensive and up-to-date coverage of the "green" process of reverse osmosis in industrial applications, completely updated in this new edition to cover all of the processes and equipment necessary to design, operate, and troubleshoot reverse osmosis systems. *NOW AVAILABLE!*

Pavement Asset Management, by Ralph Haas and W. Ronald Hudson, with Lynne Cowe Falls, ISBN 9781119038702. Written by the founders of the subject, this is the single must-have volume ever published on pavement asset management. *NOW AVAILABLE!*

Open Ended Problems: A Future Chemical Engineering Approach, by J. Patrick Abulencia and Louis Theodore, ISBN 9781118946046. Although the primary market is chemical engineers, the book covers all engineering areas so those from all disciplines will find this book useful. *NOW AVAILABLE!*